SpringerBriefs in Applied Sciences and Technology

PoliMI SpringerBriefs

Series Editors

Barbara Pernici, DEIB, Politecnico di Milano, Milano, Italy

Stefano Della Torre, DABC, Politecnico di Milano, Milano, Italy

Bianca M. Colosimo, DMEC, Politecnico di Milano, Milano, Italy

Tiziano Faravelli, DCHEM, Politecnico di Milano, Milano, Italy

Roberto Paolucci, DICA, Politecnico di Milano, Milano, Italy

Silvia Piardi, Design, Politecnico di Milano, Milano, Italy

Gabriele Pasqui, DASTU, Politecnico di Milano, Milano, Italy

Springer, in cooperation with Politecnico di Milano, publishes the PoliMI Springer-Briefs, concise summaries of cutting-edge research and practical applications across a wide spectrum of fields. Featuring compact volumes of 50 to 125 (150 as a maximum) pages, the series covers a range of contents from professional to academic in the following research areas carried out at Politecnico:

- Aerospace Engineering
- Bioengineering
- Electrical Engineering
- Energy and Nuclear Science and Technology
- Environmental and Infrastructure Engineering
- Industrial Chemistry and Chemical Engineering
- Information Technology
- Management, Economics and Industrial Engineering
- Materials Engineering
- Mathematical Models and Methods in Engineering
- Mechanical Engineering
- Structural Seismic and Geotechnical Engineering
- Built Environment and Construction Engineering
- Physics
- Design and Technologies
- Urban Planning, Design, and Policy

http://www.polimi.it

Graziano Salvalai · Enrico Quagliarini ·
Juan Diego Blanco Cadena · Gabriele Bernardini

Slow Onset Disasters

Linking Urban Built Environment
and User-oriented Strategies to Assess
and Mitigate Multiple Risks

Graziano Salvalai
Department of Architecture, Built
Environment and Construction Engineering
Politecnico di Milano
Milan, Italy

Enrico Quagliarini
Department of Construction, Civil
Engineering and Architecture
Università Politecnica delle Marche
Ancona, Italy

Juan Diego Blanco Cadena
Department of Architecture, Built
Environment and Construction Engineering
Politecnico di Milano
Milan, Italy

Gabriele Bernardini
Department of Construction, Civil
Engineering and Architecture
Università Politecnica delle Marche
Ancona, Italy

ISSN 2191-530X	ISSN 2191-5318 (electronic)
SpringerBriefs in Applied Sciences and Technology
ISSN 2282-2577	ISSN 2282-2585 (electronic)
PoliMI SpringerBriefs
ISBN 978-3-031-52092-1	ISBN 978-3-031-52093-8 (eBook)
https://doi.org/10.1007/978-3-031-52093-8

© The Author(s), under exclusive license to Springer Nature Switzerland AG 2024

This work is subject to copyright. All rights are solely and exclusively licensed by the Publisher, whether the whole or part of the material is concerned, specifically the rights of translation, reprinting, reuse of illustrations, recitation, broadcasting, reproduction on microfilms or in any other physical way, and transmission or information storage and retrieval, electronic adaptation, computer software, or by similar or dissimilar methodology now known or hereafter developed.
The use of general descriptive names, registered names, trademarks, service marks, etc. in this publication does not imply, even in the absence of a specific statement, that such names are exempt from the relevant protective laws and regulations and therefore free for general use.
The publisher, the authors, and the editors are safe to assume that the advice and information in this book are believed to be true and accurate at the date of publication. Neither the publisher nor the authors or the editors give a warranty, expressed or implied, with respect to the material contained herein or for any errors or omissions that may have been made. The publisher remains neutral with regard to jurisdictional claims in published maps and institutional affiliations.

This Springer imprint is published by the registered company Springer Nature Switzerland AG
The registered company address is: Gewerbestrasse 11, 6330 Cham, Switzerland

Paper in this product is recyclable.

Preface

This book collects the results obtained during three years of activity by a research team from Politecnico di Milano led by Professor Graziano Salvalai and the team coordinated by Prof. Enrico Quagliarini from Università Politecnica delle Marche.

The work represents the main results of the multidisciplinary activity carried out in the context of the BES^{2}ECURE project which was supported financially by the Italian Ministry of Education, University, and Research (MIUR). The book aims to provide an overview of the Slow Onset Disasters (SLOD) in the urban Built Environment (BE) discussing the strategies to assess and mitigate multiple climate change-related risks.

Climate change evidence has been reported in the last decades, suggesting that anthropogenic activities are influencing these changes towards a warmer and more polluted environment. In this context, SLODs have been linked to climate change-related disasters and have been stated to have a higher impact risk within densely built environments (i.e., cities).

Therefore, this book presents a description of the most relevant SLODs, their significance and confluence, the way in which scientists and entities are monitoring their progression at different scales, a structured risk assessment strategy, and the deconstruction of the BE characteristics that make it more prone to SLODs risk.

In addition, this book highlights the necessity of adapting the traditional risk assessment methods, to account for different vulnerability types, including the morphology and materiality of the BE, and the BE users' characteristics. In fact, individual features influence users' responses and tolerance to environmental stressors, because of age, health, gender, habits, and behaviour, thus impacting the users' vulnerability. Exposure can then amplify these issues, since it defines the number of users that can be effectively affected by the SLOD.

Starting from this perspective, the book first traces literature-based correlations between individual features, use behaviour, and individual response to the SLOD-altered open spaces. Then, a novel methodology, to quantify the variations of users' vulnerability and exposure, is offered, to support designers in quickly defining input scenarios for risk assessment and mitigation. Lastly, the book demonstrates, through a case study, the SLOD risk assessment framework proposed and the evaluation of the efficacy of risk mitigation strategies.

This book takes a practical approach and serves a large audience. It is relevant for M.Sc. and Ph.D. students but also for national and local government agencies (e.g., policymakers) and for industry. It also provides an invaluable reference for key stakeholders in the field of built environment resilience.

Milan, Italy	Prof. Graziano Salvalai
Ancona, Italy	Prof. Enrico Quagliarini
Ancona, Italy	Dott. Gabriele Bernardini
Milan, Italy	Dott. Juan Diego Blanco Cadena

Acknowledgments The work presented in the present publication has been supported by the Italian Ministry of Education, University, and Research (MIUR) Project BES^{2}ECURE—(make) Built Environment Safer in Slow and Emergency Conditions through behavioUral assessed/designed Resilient solutions (Grant number: 2017LR75XK). The authors want to thank all the project partners for their contribution to the research.

Contents

1 **SLODs in Urban Built Environment** 1
  1.1 Climate Change and Population Trends in Urbanized Area 1
  1.2 Urban Heat Island and Heat Stress 3
      1.2.1 Monitoring and Communicating the Degree of Heat Stress 5
  1.3 Urban Outdoor Air Pollution 7
      1.3.1 Monitoring Pollutant Concentrations and Communicating the Degree of Air Pollution Distress 8
  1.4 Coincidence of Arousal 9
      1.4.1 Macro Scale 9
      1.4.2 Urban Scale 13
      1.4.3 Micro Scale 15
  1.5 SLODs Risk Assessment 17
      1.5.1 Physical Vulnerability: BE Features Boosting Distress 20
  References 26

2 **User's Factors: Vulnerability and Exposure** 31
  2.1 Users' Vulnerability and Exposure: Main Definitions 31
  2.2 Individual Features 32
  2.3 Behavioural Issues in SLODs in Public Open Spaces 36
      2.3.1 Specific Movement and Behaviours 37
      2.3.2 The Role of the Built Environment: Use and Related Users' Typologies, Attractors and Repulsors 39
      2.3.3 SLODs Conditions Affecting Users' Behaviours: Air Pollution and Increasing Temperatures 41
  2.4 Users' Factors Dynamics 46
      2.4.1 Detection of Indoor and Outdoor Intended Uses 46
      2.4.2 Quick Occupant Load for Users' Exposure and Related Temporalities 50
      2.4.3 Users' Vulnerability and Related Temporalities 51

		2.4.4	KPIs for Time-Dependent Assessment	55
	References			58
3	**Quantifying SLODs Risk and Mitigation Potential in Urban BE: A Behavioural Based Approach**			**65**
	3.1	Behavioural-Based Approach Assessment of SLOD Risks in Urban BE		66
	3.2	From Scenario Creation to SLODs Simulation in Public Open Spaces or Urban BE		68
	3.3	Measuring Users' Factors: Behavioural-Based Simulations and KPIs Definition		70
		3.3.1	Users' Distribution in the Public Open Spaces	71
		3.3.2	Heat Stress and Effects on Health	72
		3.3.3	Pollution and Effects on Health	75
	3.4	Mitigation Strategies Definition and Behavioural KPI-Based Evaluation		76
	References			82
4	**Applications to Case Studies**			**89**
	4.1	Typological and Real World Scenarios for the Application of the "Behavioural-Based" Approach on SLOD Risk Assessment		90
	4.2	Time-Dependent Assessment of Users' Vulnerability and Exposure		97
	4.3	"Behavioural-Based" Assessment of SLODs Risk in the Current Scenario		100
		4.3.1	From SLOD Simulation to the Assessment of Users' Distribution in the Public Open Spaces	101
		4.3.2	Heat Stress Assessment and Effects on Health	104
		4.3.3	Air Quality Assessment and Effects on Health Due to Pollution	105
		4.3.4	Multirisk Assessment	106
	4.4	Risk Mitigation Strategies Evaluation		106
	References			112
5	**Conclusions and Perspectives**			**115**
	5.1	Urban Built Environment and Slow-Onset Disasters: How the "Behavioural-Based" Approach Could Support Risk Assessment and Mitigation?		116
		5.1.1	Physical Vulnerability	118
		5.1.2	Users' Exposure	119
		5.1.3	Users' Vulnerability and SLODs Effects from a Multi-risk Perspective	119
	5.2	Applications of the "Behavioural-Based" Approach: Insights from Case Studies and Work Perspectives		120
		5.2.1	Usability of BETs for Quick Assessment Purposes	121

	5.2.2	From Single to Multi-risk Assessment of Heat and Air Pollution	121
	5.2.3	User's Behaviour Variance in Risk	122
	5.2.4	Risk and Use of Public Open Spaces Due to Heat and Air Pollution	123
References			123
Index			125

About the Authors

Graziano Salvalai, Ph.D. and M.Sc. in Building and Architectural Engineering from Politecnico di Milano (Italy) and currently an associate professor at Politecnico di Milano, Department of Architecture, Built Environment and Construction Engineering. He has been a visiting scientist at Fraunhofer Institute for Solar Energy System in Freiburg (DE), in the Solar Building Department in 2009, and a visiting scientist at the Power System Design and Studies Group of the National Renewable Energy Laboratory (Golden—CO) in 2017. He has been a visiting professor at Colorado University at Boulder (Boulder—CO) and a guest researcher at Drexel University (Philadelphia—PA) in 2018. His research focuses on innovative building construction techniques and on the integration of building system and its control through energy simulation tools. He is a scientific director of the Building Energy Efficiency Pilot laboratory at Politecnico di Milano, and he is responsible of several scientific consultancy works in the field on energy-efficient building technologies. He is a local coordinator of four Horizon 2020 funded projects and two competitive researches (PRIN) funded by the Italian Ministry of Education, University, and Research. He is the author of three books and more than 50 Scopus-ranked scientific papers (h index = 16).

Enrico Quagliarini, Ph.D. is a full professor of Architectural Engineering and is the head of the Department of Construction and Civil Engineering and Architecture at the Università Politecnica delle Marche at Ancona (Italy). His skills are focused on the conservation and retrofitting of historic constructions and now mainly deal with the behavioural-based assessment of existing built environments for risk reduction and evacuation strategies in single and multi-hazard events such as slow onset (i.e. heatwaves and air pollution) and sudden onset disasters (i.e. earthquake, flood, overcrowding, and terrorist acts). He is a reviewer for the most important international journals on these topics, on which he attended over 50 invited lectures in national and international events. He is an author of more than 310 international and national scientific papers, including four books, 136 Scopus-ranked papers (h index = 32), and one patent. He is also a principal investigator/research staff member of several funded National and European competitive research projects. Consultant on several monuments, he participated to the Great Pompeii Project in 2015.

Juan Diego Blanco Cadena, Ph.D. and M.Sc. Engineer from Politecnico di Milano, B.Sc. Civil Engineer from Pontificia Universidad Javeriana, and currently a Postdoc Research Associate at Politecnico di Milano within the ABC Department who is currently an active member of the research Group SEEDLab@Polimi. His previous published research works have concentrated on building materials characterization, building performance simulation and monitoring analysis, building operational phase boosted with digital and sensing technologies, methods and strategies to integrate personalized visual comfort and thermal comfort monitoring into building management systems, and outdoor thermal comfort and associated health risks. He is intrigued by Building Simulation and Performance Analysis and is devoted to the study of human-building interaction boosted by digital technology. He is eager to achieve sustainable and smart building design and operation by making them able to adapt to the surrounding climate and to the dynamic occupants' comfort demands. In addition, the implementation of emerging technologies, programming, scripting, and data science applied to building facade components are the core of the examined research areas.

Gabriele Bernardini, Ph.D. is a researcher and lecturer at the Università Politecnica delle Marche at Ancona (Italy). Formed as building engineer-architecture, he is interested in developing tools for including the users' factors in building design, operation, and maintenance. In this context, his work is widely focused on users' safety in the built environment (i.e. emergency and evacuation), by involving the building, the open space, and the urban scale. He develops methods for the analysis of users' exposure behaviours in case of different emergencies, simulation models and software for evacuation analyses (e.g.: heatwaves, air pollution, climate change, fire, earthquake, flooding), and innovative systems for increasing users' awareness before a disaster and helping people during an emergency in indoor and outdoor built environment. He has published many papers in international journals and international conferences concerning the advances of our research group works. He is an author of more than 110 publications, having national and international relevance, most of them focused on users' safety in emergencies, risk assessment and mitigation, including 90 Scopus-indexed documents (h index = 20) and one SpringerBrief book.

Chapter 1
SLODs in Urban Built Environment

Abstract Climate change evidence has been reported in the last decades, suggesting that the anthropogenic activities are accelerating these changes towards a warmer and more polluted environment. In addition, it has been noted that cities are the areas that have generated a larger disturbance, and which will probably suffer the most from these two aspects. In this context, Slow-Onset Disasters (SLODs) have been linked to climate-change related disasters and have been stated to have a higher impact risk within dense Built Environment (BE) (i.e., cities). Therefore, this chapter presents a description of the most relevant SLODs, their significance and confluence, the way in which scientists and entities are monitoring their progression at different scales, a structured risk assessment strategy and the deconstruction of the BE characteristics that make it more prone to SLODs risk. Finally, this chapter highlights the necessity of adapting the traditional risk assessment methods, to account for different vulnerability types (physical of the BE and of the exposed citizens).

Keywords Climate change · Slow-onset disasters · Built environment · Urban heat island · Air pollution

1.1 Climate Change and Population Trends in Urbanized Area

The urban Built Environment (BE) has always been of interest for inhabitants, researchers, and designers. In fact, it has become the main preferred living location for most of the world population. To be precise, the United Nations [1] has reported in 2018 that 55% of the world's population was already living in highly urbanized areas and that this ratio is projected to grow up to 68% by 2050 [2]. As a result, the understanding of its complexity and dynamics is becoming highly relevant to support designers, policy makers and researchers. At the same time, this condition motivates them to scrutinize and establish potential development solutions that improve the quality of life of the users of such anthropogenic environment.

These urban population growth projections, coupled with the projected health threats identified by the World Health Organization (WHO) [3, 4] due to climate change, foresee that around 6.7 billion people are at great risk. Thus, identifying climate change related risks within the urban environment has become a pillar in the Sustainable Development Goals (SDGs); in specific, the following three topics: good health and wellbeing, sustainable cities and communities, and climate action [5].

To start with, climate change related risks can be associated with the definition of Slow Onset Disaster risks (SLODs). Given their slow unfolding. Or slow development, but substantial economic, social, and environmental impact as described by Gunn and Noji [6, 7]:

> … vast ecological breakdown in the relation between humans and their environment, a serious and slow event on such a scale that the stricken community needs extraordinary efforts to cope with and resolve it …

In addition, the urban BE has demonstrated to have an inherent mesoscale environment and higher dynamism, compared to the rural areas (normally performing worst in terms of climate change attenuation). But also, within the same urban BE, quadrants would behave differently, environmentally-wise, according to their characteristics resulting in diverse micro-climates [8–12]. And those worse urban BE performant microclimates are strongly linked to the most frequent and critical climate-related SLODs: increasing air pollution and temperatures [13].

These SLODs be considered as such given that increasing air pollution represents a risk for humanity, and for the natural environment when the contaminants production rate surpasses the contaminants assimilation rate. This would promote land and forest degradation, and some species would not endure the quality of the air they are breathing, then biodiversity loss would be inevitable. Consequently, the ocean would try to balance the contaminants cycle by absorbing more contaminants at a pace faster than it can process developing its own acidification [14]. Finally, some of these contaminants are among those considered Green House Gases worsening the heat entrapment in the atmosphere which contributes to the increasing temperature trend. On the other hand, increasing temperatures enhance water vapor carrying capacity, moving onto water vapor state what should have been used for land moisture. This could affect the process of desertification, droughts, and forest degradation. Thus, loss of biodiversity is expected. In addition, higher amount of water vapor in the air would generate a larger risk for extreme precipitation events in the eventuality of a pressure and/or temperature drop. Higher temperatures around the globe would also boost the glacial melting making more evident the glacial retreat, and, in consequence, rising the sea level and decreasing the availability of freshwater can be expected [15].

Therefore, SLODs are likely to occur in synergy and evidence events related to them are probable to arise in parallel or one after the other, and thus, monitor their evolvement is essential to better plan mitigation and adaptation measures. To this end, based on the findings reported by Salvalai et al. on SLODs interaction [16] and projected rapid urbanization rate [12, 17, 18], increasing air pollution and temperatures (and the Urban Heat Island—UHI as correlated phenomena) are hereby studied in depth.

1.2 Urban Heat Island and Heat Stress

UHI has been defined as heat accumulation phenomenon, characterized by further sensed air temperature increase, that develops in urban areas due to constructions and human being activities. Several urban and suburban areas experience higher temperatures compared to the closest rural surrounding areas [19]. According to EPA [19], UHI (Fig. 1.1) can be divided and studied in two different domains:

- Surface Heat Island (SHI);
- Atmospheric Heat Island (AHI).

The SHI is referred to the effect of the direct sun that heats the exposed built surfaces like roofs, pavement to temperatures hotter than the air [20]. SHI reaches the maximum level during the day and can prevail during the night depending on the thermal inertia of the built surfaces (i.e., thermal mass). The intensity changes according to the season as well as the ground cover (extension and type). Direct and indirect methods can be used to identify the SHI, and its intensity. For instance, remote sensing for estimating surface temperatures is often employed to better understand the

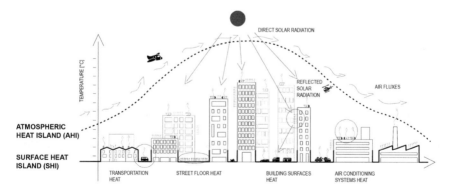

Fig. 1.1 Schematic UHI representation, including the atmospheric and surface heat island within the BE

Table 1.1 Description of the SLOD incremented risk by the UHI phenomenon

Urban heat island actors						
Causes	Anthropogenic heat	Urban geometry	Properties of urban materials	Lack of greenery	Weather	Geographical location
Actors	Automobiles, buildings, industries, and air condition units	Urban morphology modifies the solar radiation exposure	Opaque surfaces with high absorptivity (e.g., asphalt, concrete, black surfaces)	Decreased evaporative heat losses and lower shading potential	Clear sky, calm wind, and excessive solar radiation	Proximity to large water bodies or mountains

phenomena (e.g., infrared images). For this, different institutions (e.g., Copernicus,[1] NASA[2]) offer the possibility of accessing recent and historical results derived from satellite readings. Unfortunately, they are limited to the space and time resolution, as well as bounded to the moment in time in which the satellite is over the place of interest [21].

Instead, the AHI represents the differences between the air temperature of the urban area and the air temperature of the surrounding area. The air temperature is intended as the air temperature of the space where the pedestrian and residents coexist, that is, from the ground level to the top of the trees and roofs. The AHI intensity fluctuations are lower than the SHI, and thus, to be able to observe and address this phenomenon, a dense sensor network infrastructure should be put in place. Once again, different regional institutions make their data available from the weather station network owned and maintained by them (e.g., ARPA in Italy[3]) or from interpolated models (e.g., ERA5-Copernicus[4]) for the area of interest.

Figure 1.1 traces a schematic overview of which "actors" are involved in the heat transfer dynamics of AHI and SHI for an easier understanding of their interaction. Both AHI and SHI contribute to the consolidation of the UHI effect, which is worsened by the evident and reported globally gradual temperature increase SLOD (i.e., potentially > 1.5 °C) [22, 23]; a detailed explanation on how the UHI depicted in Fig. 1.1 arises is presented in Table 1.1.

UHI can be considered one of the more severe threats to the health of the city population resulting from climate change, as heat wave events are becoming increasingly more frequent and intense. For instance, the increasing value of death per heat wave events in the last decade confirms the criticality of the trend of temperature

---

[1] https://cds.climate.copernicus.eu/ (24/08/2023).
[2] https://modis.gsfc.nasa.gov/data/dataprod/mod11.php (13/09/2023).
[3] https://www.mase.gov.it/pagina/agenzie-regionali-protezioni-ambiente (13/09/2023).
[4] https://cds.climate.copernicus.eu/cdsapp#!/dataset/reanalysis-era5-single-levels?tab=overview (13/09/2023).

1.2 Urban Heat Island and Heat Stress

Table 1.2 Meteorological disasters summary from 1900 to 2019 (extracted from [25])

Continent	Date range	Events count	Total deaths	Deaths/event	Total affected	Affected/event
Africa						
	1900–2019	8	291	36	86	11
	2000–2019	6	237	40	86	14
	2010–2019	3	137	46	86	29
Americas						
	1900–2019	35	6177	176	20,221	578
	2000–2019	11	591	54	17,521	1593
	2010–2019	5	265	53	17,490	3498
Asia						
	1900–2019	78	16,860	216	299,323	3837
	2000–2019	50	9825	197	298,583	5972
	2010–2019	24	6224	259	290,522	12,105
Europe						
	1900–2019	75	138,566	1848	2120	28
	2000–2019	63	137,319	2180	1850	29
	2010–2019	17	60,213	3542	–	–

rise (Table 1.2). This phenomenon may affect more disadvantaged categories as low-income classes (more exposed to the outdoor built environment), elderly, children and people already affected by health conditions (low body thermoregulation capacity). In fact, in 2014, WHO [3] estimates that by 2030, more than 92,000 additional heat related deaths are expected worldwide; and, more than 255,000 for 2050 if no measures are taken. Lee and Kim [24] forecasted, in the context of South Korea, for 2090s, a ×4 or ×6 increase in temperature-related mortality with respect to the 1992–2010 period.

1.2.1 Monitoring and Communicating the Degree of Heat Stress

Cities around the world have installed several weather stations able to capture trends of environmental conditions related to thermal perception and heat stress. They normally register and openly store the time history of levels of air temperature, relative humidity, wind speed and horizontal irradiation. These stations are normally used alone or combined to allocate a thermal stress rating to communicate the condition of thermal stress [26].

The most widely used stations are composed of sensors to monitor and communicate, or compute:

- Air temperature—considers only the overall atmospheric temperature.
- Heat Index (HI)—considers temperature and humidity levels [27].
- Temperature Humidity Wind index (THW) index—considers air temperature, humidity, and the wind cooling effect.[5]
- Temperature Humidity Solar and Wind (THSW) index—considers the air temperature, humidity, the wind cooling effect, and the solar radiation heating effect.[6]
- Mean Radiant Temperature (MRT)—considers air temperature, humidity, wind cooling effect, and the radiative heat coming from surrounding surface [28].

Other metrics can be used to consider the perception of thermal stress considering a particular body type and functioning [29]. These allow to capture a more realistic value of perceived thermal stress. However, it is complex to establish the variability of the perception given the context in which they are tested. Such metrics can be based on [26]:

- Outdoor Standard Effective Temperature (OUT_SET) [26]—considers heat storage and the body internal heat exchange.
- Universal Thermal Climate Index (UTCI) [30, 31]—considers heat storage and multi-node internal body heat exchange.
- Wet Bulb Globe Temperature (WBGT) [32]—considers adaptability and previous exposure to high heat stress, and has a lower sensitivity to shortwave radiation exposure [33].

Nevertheless, most of the previous computed metrics are impractical for large scale analysis, given that most of the stations do not have the direct and diffuse radiation components, or the globe temperature for accurately estimate MRT. Researchers have proposed empirical methods to compute WBGT from meteorological data and multiple regression, obtaining decent confidence and accuracy (~ 95% and ± 2 °C) [34]. Nevertheless, these methods have not yet been agreed and inserted into standards. Other studies have proposed alternative indexes that can be equally applied as the ones mentioned above, such as the ones reviewed and summarized in Blazejczyk et al. [26]; and as RiskT proposed by Blanco Cadena et al. [35]. The latter is meant to communicate the degree of thermal stress based on the superimposition of weights attributed to the arousal of undesirable conditions for each of the relevant parameters assessed in thermal stress (i.e., air temperature, wind speed, solar irradiation, and relative humidity).

---

[5] https://www.davisinstruments.com/pages/what-is-thw-index (17/04/2023).

[6] https://www.davisinstruments.com/pages/what-is-thsw-index (17/04/2023).

1.3 Urban Outdoor Air Pollution

Urban outdoor air pollution refers to the excess and concentration of air pollutants experienced by those living in and around urban areas (i.e., cities), influenced by the emissions of industry, traffic, and buildings' plant systems. For instance, the latter emissions represent the 39% of global energy-related carbon emissions and are distributed as follows, and summarized in Table 1.3 [36, 37]:

- 28% related to buildings in operation: heating, cooling, and lighting. Energy use is heavily impacted by the quality of building envelopes, with emissions especially substantial in older building stock.
- 11% related to embodied carbon in the construction process: waste generation, water use, dust creation and greenhouse gas emissions.

This pollutant concentrations have direct effects in nature and in humans. But not only, they have been proven to potentiate the effect of UHI by creating a heat outwards impermeable layer in the atmosphere [38] (Fig. 1.2). Therefore, actions should be directed to tackle both in parallel to make sure the mitigation strategies are both effective and not counterproductive for the other phenomenon.

According to WHO, air pollution represents one of the biggest environmental risks to health. That is, 91% of the world's population lives in a place with poor air quality. Also, only 20% of the urban population have been reported to live in areas complying with the suggested healthy air pollutant concentration limit thresholds for pollutants such as the Particulate Matter sized 2.5 μm (PM2.5) [39, 40].

Table 1.3 SLOD urban air pollution description

Urban outdoor air pollution						
Causes	Anthropogenic processes	Urban geometry	Properties of urban materials	Lack of greenery	Weather	Geographical location
Actors	Automobiles, buildings, industries, and air condition units	Building, road, sidewalks, and courtyard morphology, wind	Opaque surfaces: asphalt, concrete, black surfaces	Decreased particulate matter absorption	Clouds with high atmospheric pressure, wind intensity, wind direction	Proximity to three field

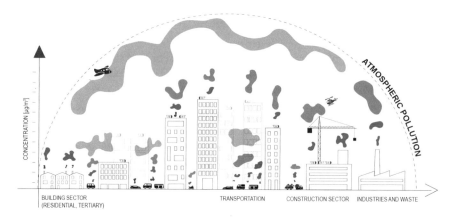

Fig. 1.2 Schematic air pollution representation, including the impact on atmospheric heat island within the BE

1.3.1 Monitoring Pollutant Concentrations and Communicating the Degree of Air Pollution Distress

In the same fashion as for weather conditions, cities have installed dispersed air quality stations to capture, monitor and alert about the trends of air pollutant concentrations. These stations normally register and store the time history of concentration levels of carbon, nitrous oxide, ozone, particulate matter, and/or volatile organic compounds. These can be and are normally used, alone or combined, to compute an indicator to establish the current level of quality of the air that citizens breathe. In this case, one common indicator has been stablished around the globe, however with different quality levels or scales:

- Air Quality Index (AQI)—calculated for every pollutant type, based on a relative evaluation of how distant the current concentration is, compared to established limit thresholds for each level [41].

Such established threshold limits are normally given by the health authorities and can be based on the indications given by WHO [42]. There are no established correlations between the values of AQI and health burden, to the authors knowledge, except for the adjusted AQI into Air Quality Health Index (AQHI) work presented by Cao et al. [43] to predict the risk of mortality. The more direct relationship remains on the use of the relative risk allocation given the time of exposure to certain air pollutant concentrations, following a similar process as the one presented by Blanco Cadena et al. [44] applied to a specific climate and within a representative urban BE.

1.4 Coincidence of Arousal

Given the nature of the critical SLODs studied, it is physically possible and probable that their unfolding and disaster impact arise consecutively or in parallel. As introduced in previous sections, scientists, organizations, and institutions have presented proof on how these two SLODs hazard is progressively becoming more intense and their related events more frequent. In other words, it is feasible that the events that evidence the presence and intensification of such SLODs (i.e., UHI and air pollution). In fact, now a days cities are more frequently under bad air quality and suffering longer and more intense warm temperatures [45–47] ($\times 2.7$ times in this century with respect to the last century, and $\times 1.4$ comparing the last decade with the last century [25]). Monitoring campaigns and stored time series have allowed researchers to identify and better plan strategies to mitigate such risks individually, but few have studied and tackled the joint challenge. In the following sections some examples are presented.

1.4.1 Macro Scale

As mentioned in Sects. 1.2.1 and 1.3.1, there are diffused weather and air quality monitoring networks. These networks allow institutions and organizations to keep track of climatological and environmental trends, but also to establish if the enacted, and/or promoted, mitigation/adaptation actions and policies are having a significant impact. For instance, The National Oceanic and Atmospheric Administration (NOOA)[7] governmental agency collects, analyzes, and reports global ocean and land surface temperatures since 1880s. And thus, it can provide insights on the way that world-wide temperatures are rising at a rapid pace (Fig. 1.3). In addition, open databases, like the one hosted by the World Bank financial institution,[8] can provide information on global development trends, including environmental development. For the latter, it is possible to understand how CO_2 emissions (most common utilized metric to correlate air pollution to anthropogenic processes) have grown at a similar rate than the temperatures (Fig. 1.4).

Such trends have been associated with the increase of climate-related reported and documented disasters (~ $\times 2.7$ times more fog and heat waves arousal than last century) characterized by of a growing frequency, intensity and duration. In fact, longer heat waves around the world are expected with the proposed scenarios of climate based on the potential future of carbon emissions (e.g., representative concentration pathways (RCPs)) [48]. An indication can be found in Table 1.2, and Fig. 1.5. The latter maps foreseen the growing trend of heat wavelength (in days) in Europe based on bias adjusted output from the EURO-CORDEX[9] ensemble of

---

[7] https://www.noaa.gov/ (04/05/2023).
[8] https://data.worldbank.org/ (04/05/2023).
[9] https://cds.climate.copernicus.eu/ (24/08/2023).

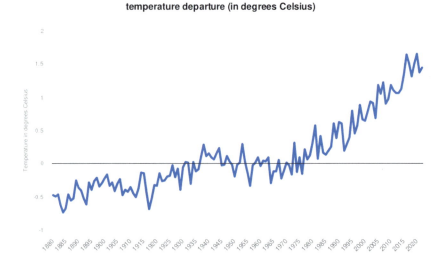

Fig. 1.3 National Oceanic and Atmospheric Administration (April 3, 2023). Annual anomalies in global land surface temperature from 1880 to 2022, based on temperature departure (in degrees Celsius) [graph]. In Statista. Retrieved October 30, 2023, from https://www.statista.com/statistics/1048518/average-land-sea-temperature-anomaly-since-1850/

climate models. The maps show the forecasted results for two representative Concentration Pathways: (1) RCP 4.5, defined as intermediate scenario (global temperature rise between 2.5 and 3 °C by 2100, due to global CO_2 eq concentrations at ~ 650 ppm); and (2) RCP 8.5 as a pessimistic scenario (global temperature rise of 5 °C by 2100, due to global CO_2 eq concentrations at ~ 1370 ppm). The projection for both the scenarios shows a strong increase (~ ×2) of the phenomena with more than 6 and 8 days in average (9 and 11 days at the 99th percentile) for 2050, respectively RCP 4.5 and RCP 8.5; that could increase to 8 and 24 in average (11 and 31 days at the 99th percentile) for 2080. Their frequency, intensity and length are thus expected to increase, leading to a substantial increase in mortality, especially in vulnerable populations.

1.4 Coincidence of Arousal   11

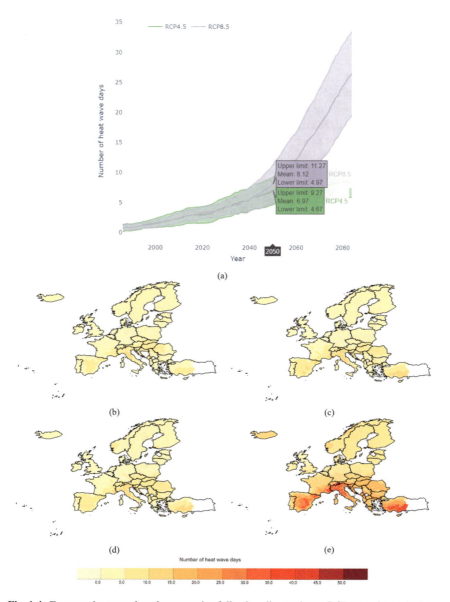

Fig. 1.4 European heat wavelength progression following climate change RCP scenarios and **a** time series trend, elaborated with the online exploratory tool for the heat waves and cold spells in Europe derived from climate projections dataset.[10] Starting from **b** 2050 with RCP 4.5, **c** 2050 with RCP 8.5, **d** 2080 with RCP 4.5 and **e** 2080 with RCP 8.5. Considering a heat wave if at least three consecutive days the daily maximal temperature exceeds the 99th percentile of the control period (1971–2000)

---

[10] https://cds.climate.copernicus.eu/apps/c3s/app-health-heat-waves-projections?Definition=Euroheat%20project (14/09/2023).

12   1   SLODs in Urban Built Environment

To confirm the criticality of the potential future scenarios, Fig. 1.5 shows the trend in heat attributable mortality incidence (annual deaths/million/decade) for the general population, 2000–2020; and Fig. 1.6 displays the PM2.5 attributable mortality incidence in 2020 (source European Environment Agency (EEA)[11]).

From this results, it is possible to note how citizens in eastern Europe, south of Spain and north-west of Italy have suffered the most reported heat and air pollution linked deceased citizens in the past years. Thus, in a yearly average, the mentioned regions have suffered more than 40 heat attributable deaths per million in average. And, they have reported in 2020 between 50 and 150 PM2.5 attributable deaths per

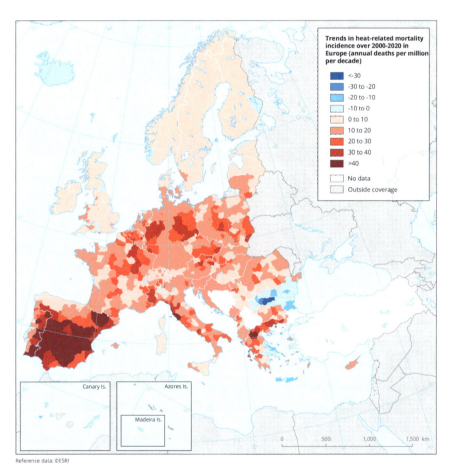

Fig. 1.5 The map shows the trend in heat attributable mortality incidence (annual deaths per million per decade) for the general population, 2000–2020. *Source* EEA[12]

---

[11] https://www.eea.europa.eu/ (24/08/2023).

[12] https://www.eea.europa.eu/data-and-maps/figures/trends-in-heat-related-mortality   (24/08/2023).

1.4 Coincidence of Arousal

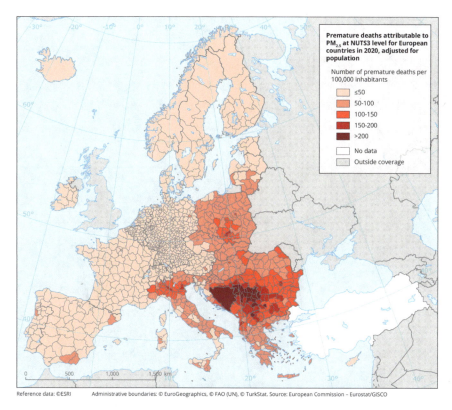

Fig. 1.6 The map shows the PM2.5 attributable mortality incidence in 2020 (annual deaths per million per decade) for the general population. *Source* EEA[13]

100,000 inhabitants. Hence, motivating the public and private bodies to act upon mitigating the current evolution of the aforementioned SLODs to guarantee safety to citizens.

1.4.2 Urban Scale

Likewise, effects at a more granular space resolution are available, discerning between countries and national regions. However, these result analyses are also available at the urban scale, to better target risk mitigation assessments, as cities have their own resources and characteristics. That is the case for the municipality of Milan, individuated by Salvalai et al. [49] as a city of interest for temperature- and pollution-related SLOD risk analysis (based on risk hazard intensity and risk

---

[13] https://www.eea.europa.eu/data-and-maps/figures/premature-deaths-attributed-to-pm2-1 (24/08/2023).

14	1 SLODs in Urban Built Environment

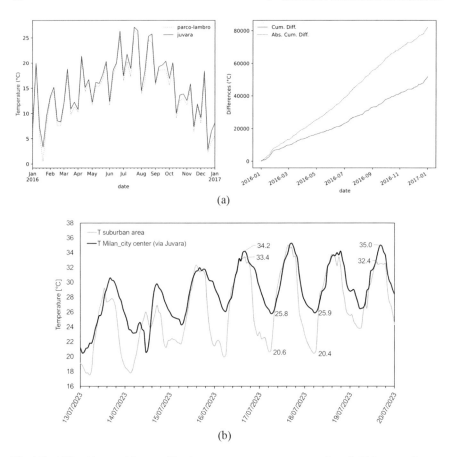

Fig. 1.7 AHI evidence with **a** weekly air temperature average comparison (left) between Parco-Lambro (sub-urban and large greenery coverage) and Juvara (urban and low greenery coverage) in 2016, and their hourly difference (right). And, **b** its prevalence during 13–10 July 2023, using time series gathered from ARPA Lombardia data repository[14]

exposure). At this location, weather and air quality data can be found stored from weather stations placed in sub-urban areas with larger green area coverage, and in densely urban areas, allowing a direct monitoring of AHI. That is the case for Parco-Lambro (sub-urban) and Juvara street (urban) weather station respectively. For 2016, their environmental differences can be noted in Fig. 1.7a, where air temperatures have been plotted for the whole year, together with the cumulative and absolute cumulative degree differences.

It is visible how the urban (Juvara) readings are prevalently higher than the sub-urban (Parco-Lambro). It reached an approximate daily median difference of 1 °C and a maximum daily average difference of 4.6 °C (urban vs. sub-urban). This significant air temperature differences, and AHI, were verified also in subsequent years

---

[14] https://www.arpalombardia.it/ (20/09/2023).

1.4 Coincidence of Arousal

(Fig. 1.7b—daytime differences of 1–3 °C and night time by 4–6 °C) giving strong evidence of UHI in Milan. But to understand the potential coincidence of arousal, the air temperature and air pollutants concentrations trends were plotted for 2019 to understand their future behavior in Fig. 1.8. It is worth noticing that for the case in the municipality of Milan at the urban context, there is a resemblance on the air temperature and the ozone (O_3) concentration, as well as the particulate matter (PM2.5 and PM10), with respect to the potential use of heating and cooling conditioning systems (Fig. 1.8a, b).

Moreover, on July, contemporarily daily air temperatures beyond 30 °C where recorded while O_3 concentrations were over 120 $\mu g/m^3$ (healthy limit threshold of 8 h exposure set by WHO). Likewise, for the same period, the particulate matter (PM2.5 and PM10) are dangerously close to the healthy limit thresholds (25 and 50 $\mu g/m^3$ respectively).

These results confirm that in urban areas, given recurrent negative impact, the unleashing of increasing temperatures and air pollution SLODs in parallel is possible. But, also that their evidence-events (heat waves and smog) could superimpose, in terms of days or weeks in the future.

1.4.3 Micro Scale

As mentioned earlier, there is a low number of published works on mitigating strategies or action plans that conjunctly tackle increasing temperatures and air pollution. Even though at a smaller scale input data should be less limited, processing should be of lower complexity, computational cost, and others [29, 50]. This lack of published works could be related to the effort that is required to make a survey campaign with validated sensor units, and to maintain the monitoring campaign for longer periods.

In fact, Takebayashi and Moriyama presented a series of case studies in Japan of heat adaption measures that present results that lasted only for a couple of hours in a day or for a week [51], and without air pollution depletion analysis. Likewise, Abhijith et al. presented a review of different short-term analysis mainly referred to computer-based simulated assessments [52]. And Kukkonen et al. a series of analysis on short-term measured smog events worldwide, highlighting their current and future criticalities [53].

Also, Rampini et al. collected and mapped outdoor air pollutants using a portable-low cost device along some streets [54]. The campaign was carried out on a portion of the area discussed in the previous section, in the proximity of M. Pascal Città Studi air quality station (< 550 m). The obtained results, suggest that the values of particulate matter can reach dangerous concentrations at lower space density during certain seasons (Fig. 1.9). Contrary to what large or urban scale analysis would normally imply, and alike to the thermal stress generated by increasing temperatures.

The noticeable difference in small patches in temperature and air quality has been associated with how well the surroundings shield from, absorb, and hold, heat and/or

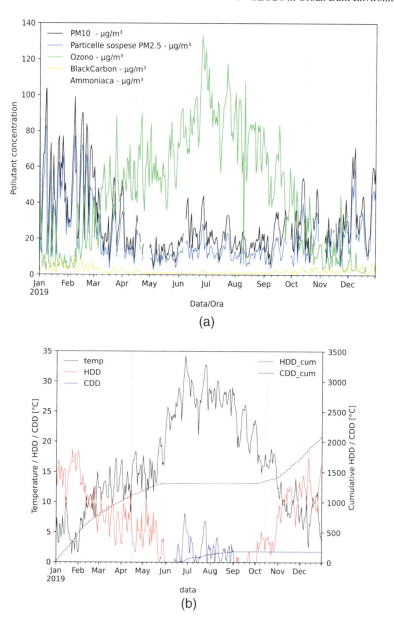

Fig. 1.8 Daily average of **a** pollutant concentration from M. Pascal Città Studi air quality station and **b** air temperature from M. Juvara weather station for 2019, calculating also daily heating and cooling degree day, and the cumulative degree days. Grey vertical lines have been drawn to show the beginning (right, Oct. 15) and end (left, Apr. 15) of the municipality's heating season

1.5 SLODs Risk Assessment

Fig. 1.9 Survey campaign results for **a** PM2.5 and **b** PM10 on the proximity of Juvara weather station. Extracted and edited from [54]

pollutants, and how smoothly wind runs for heat alleviation and pollutant dispersion [53, 55].

1.5 SLODs Risk Assessment

As described before, SLODs have a significantly different behavior compared to any other type of risks; they develop in a diverse timeframe, thus intensity and duration. In fact, SLODs are characterized by a low intensity and lengthy, or recurrent, exposure to adverse health conditions. However, these adverse health conditions are progressively more frequent (revise trends exposed in Sects. 1.2 and 1.3), reaching the point in which, in certain geographic areas, they have become regular. Hence, to facilitate the analysis and enhance the efficacy of mitigation and adaption measures, the decomposition of the risk elements for SLODs can be modified as described in Fig. 1.10 and hereby described.

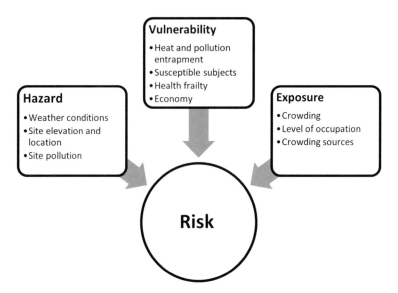

Fig. 1.10 Risk definition for SLODs, specifying its components (hazard, vulnerability, and exposure)

- Hazard—related to the adverse environmental conditions, these shall not be considered as isolated (limited risk would be identified) but as the potential conditions during a specific season. It will be evaluated in terms of the intensity and frequency of the adverse conditions.
- Vulnerability—given that environmental conditions (e.g. weather or air pollution) act on a macro scale and their variations are slow, the vulnerability can be associated to two sub-categories instead:
  (a) one linked to how the surrounding context worsens the adverse conditions (Physical Vulnerability); and,
  (b) one associated with health frailty of the person when exposed to the hazard (Social Vulnerability).
- Exposure—related to the level of occupation density of a certain space, the larger the number of citizens involved in the hazard, the larger the risk. And also, on the frequency of exposure (time-based).

Hazard has already been extensively discussed on Sects. 1.2 and 1.3, while Vulnerability and exposure will be treated in the coming sections. In particular, physical vulnerability is assessed in Sect. 1.5.1, while social vulnerability and exposure in Chap. 2.

In brief, the risk assessment of SLODs, and more in specific of increasing temperatures and air pollution, requires a multidomain analysis that studies in detail the characteristics of the environmental context, the built environment properties, and

1.5 SLODs Risk Assessment

the population features. These will determine the heat and air pollutant concentration dynamics within the built environment, thus influencing the stress to which the population is subjected to, producing short- and long-term effects on the health of citizens. An schematic summary of the actors within the SLODs risk assessment have been included in Figs. 1.10 and 1.11.

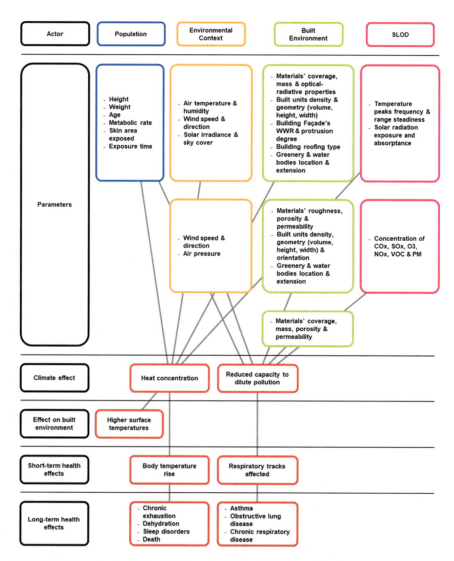

Fig. 1.11 Schematic description of the force's relationship involved in SLODs development, which might favor their risk, including the potential effects on the climate, the BE and the health of the BE's population

1.5.1 Physical Vulnerability: BE Features Boosting Distress

In detail, urban environments are not formed by the individual landscape elements that make up the street; instead, they are formed by the mutual connections of multiple elements. For example, trees and green spaces, low building layouts, the aesthetic facades of streets and buildings, and open spaces function together to amplify the attractiveness of the space [56, 57].

Urban areas, and therefore pedestrians, are particularly vulnerable and exposed to pollutants, as the urban form alters the microclimate and air quality [58], discouraging people from walking in cities and exposing them to risks depending on their individual characteristics (for example, age, health, ability). The condition is aggravated by the fact that pollutant sources are mainly concentrated in urbanized areas, such as vehicular traffic, industrial activities, heating systems, and commercial areas [57, 59]. High pollutant emissions can cause severe damage to human health [60], and variations of its concentration levels below the urban canopy could indicate the existence of specific, local, and contemporaneous anthropogenic sources that may be threatening the outdoor environmental quality [61, 62]. Furthermore, cities are affected by the well-known phenomenon of UHI [63, 64] due to their morphological peculiarities, land surface cover and usage, and lack of greenery. High UHI values compromise citizens' everyday commuting, open-air activities, and well-being in general [59]. Such behavioral modifications, and more, are further assessed and found within Chap. 2.

Moreover, in this work, strategies to evaluate the intensity of the hazard, the potential people's exposure and the boosted risk due to vulnerability, in relation with the effects on human health, have been laid down and will be further discussed in Chaps. 2 and 3. In general, dense urban areas increase the risk of the two above mentioned SLODs for the following main reasons: (1) reduced evaporation, transpiration and shading, due to limited green areas and disadvantageous geometry; (2) increased surfaces temperatures with high thermal capacity and/or low albedo; and, (3) increased air stagnation due to diminished wind speed [8–12]. Thus, making every portion of the city (e.g. neighborhood) react/provide a different meso-climate to the people within.

1.5.1.1 Geometric and Space Characteristics of the BE

Urbanization has a high impact on the microclimate of cities. As mentioned in the previous sections, structures such as buildings, roads, and other infrastructure absorb and re-emit the sun's heat more than natural landscapes such as green areas and water bodies. Thus, urban geometry, building density, and vegetation patterns and materiality of the urban fabric have relevant impacts on the UHI intensity (Fig. 1.12).

The urban geometry is formed from urban canyons that are defined by the buildings' heights and streets' widths ratio (H/W ratio) and the orientation of their long axis (O). These two parameters (closely correlated to sky view factor) drive the solar

1.5 SLODs Risk Assessment

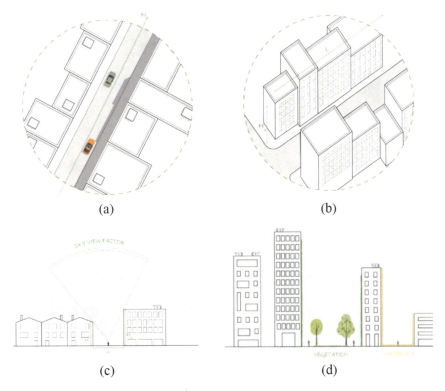

Fig. 1.12 Urban geometric and space characteristics largely influence the UHI and AP phenomena. **a** Orientation, **b** density, **c** sky view factor, **d** vegetation and materials

radiation access and wind flow. Hence, also the resulting heat reflection, absorption, and emission determined by the constructive properties (Sect. 1.5.1.2). Which, all together, influence the surface and the ambient air temperature and thus the urban heat island effect. Figure 1.13 reports a study conducted on a small portion of the Milan-Lambrate district where different streets canyons, expressed by aspect ratio and including their vegetation, are simplified in section to test and show the high difference of solar access. Such direct solar radiation analysis on the canyon surfaces considered different hours of the day is presented in Fig. 1.14. As expected, different canyon aspect ratio (H/W) and different street configuration (orientation, building height variation and presence of trees) influences the quantity of solar energy received, hence the surfaces influencing the perceived ambient air temperature.

As shown in Fig. 1.14, a higher H/W ratio means less sky view and more shading. Likewise, higher trees' canopy area. Thus, higher H/W and tree presence, lower solar radiation which can reduce the temperature in the street canyon.

Nevertheless, a higher H/W ratio can also create more complex wind patterns and vortex flows, which can affect the dispersion of pollutants and the ventilation of the street canyon [65]. Worsen by the additional drag force that entails the presence of

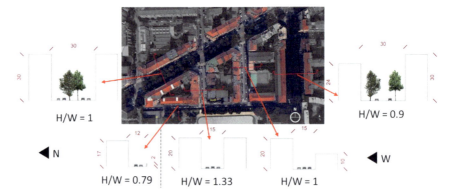

Fig. 1.13 Geometrical analysis of two different street canyon configuration in Lambrate Milano district. On the right via G. Pacini with H/W = 1 and on the left via XXX

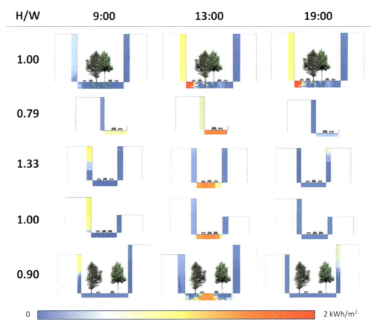

Fig. 1.14 Solar radiation analysis on different street canyon configuration of a portion of Lambrate Milano district. The irradiation intensity exposure has been calculated during June 21st for 9:00, 13:00 and 19:00

trees within the canyon. In fact, Salvalai et al. presented how the presence of trees at the wrong location, worsens the thermal perception (reported as physiological equivalent temperature—PET) and the NO_3 air concentration distribution within a single street canyon H/W ratio modelled in Milan (IT) with ENVI-met [66] (Fig. 1.15).

1.5 SLODs Risk Assessment

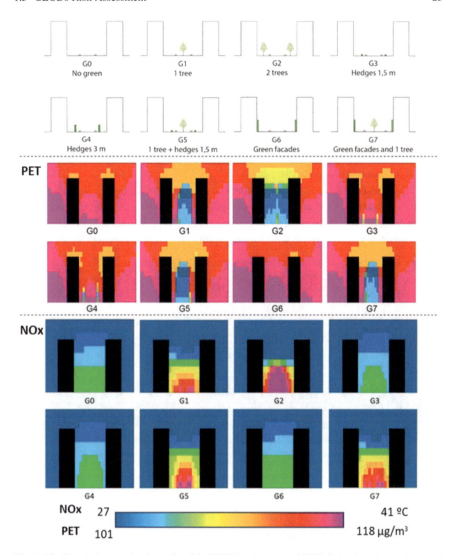

Fig. 1.15 Simulation results for a simplified H/W = 1 canyon, highlighting location and type of greenery within the section of the canyon, and the distribution in section of a thermal comfort index (PET) and the air pollution concentration of NO_x at 14:00 (extracted and edited from [66])

1.5.1.2 Constructive Properties

The UHI effect is highly affected by streets and buildings finishing. Construction materials such as concrete and asphalt can absorb heat during the day and radiate it back at night, much more than areas covered with vegetation do. The normal construction materials are great absorbers of infrared and near infrared solar radiation. As the urban areas continue to extend, they have accumulated an excess potential for heat absorption which has put them out of balance compared to more rural areas. It is therefore clear that the type of material and its radiative properties (in particular, albedo) play a role in regulating the surface temperature. The term "albedo" describes the proportion of incident radiation reflected by a system. A perfect reflector would have an albedo of 1, whereas a perfect absorber would have an albedo of 0. Depending on the types and structures of the pavement/covering materials, which can have effects on the albedo and surface heating, ground surfaces reflect solar radiation or heat the air above them. Surface that can absorb more solar radiation may turn it into heat, thus warming the air, which in turn heats urban areas and make them uncomfortable for human beings (Figs. 1.16 and 1.17).

Considering also that albedo is sometimes referred to as the proportion of solar radiation reflected by the earth back into space, it becomes an important property in the built environment for heat management. However, just like for the test facility presented in Fig. 1.17, in the built environment, there is a heterogeneity on surface

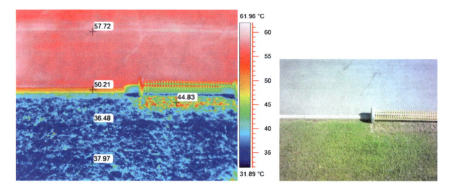

Fig. 1.16 Infrared and visible picture of a concrete and grass horizontal outdoor surface (taken in Milan (IT) on 22/08/2023 at 15:30 at ambient air temperature of 38 °C)

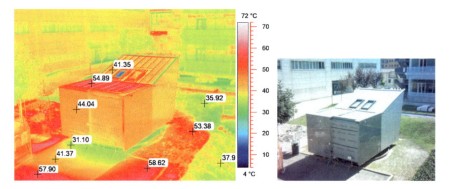

Fig. 1.17 Surface temperature variation on a test facility with different material finishing (taken in Milan (IT) on 22/08/2023 at 15:30 at ambient air temperature of 38 °C)

1.5 SLODs Risk Assessment

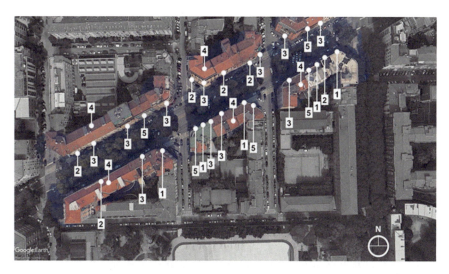

Fig. 1.18 Schematic representation of building materials along the road for via G. Pacini. *1* Marble; *2* stone cladding; *3* plaster; *4* clay roof tiles; *5* bricks; *6* ceramic tiles

materials and radiative properties, which makes it harder to propose a one-solution-fits-all. In fact, Fig. 1.18 shows an example of maps reporting the building walls finishing materials along a road in the analyzed portion of Lambrate Milano district.

Therefore, making large scale analysis on heat stress becomes challenging. In addition, surface materials and greenery have a pollutant deposition capacity that is influenced by (1) particle resuspension rate, (2) deposition velocity and (3) surface resistance (like roughness). It requires the combination of air pollutants dispersion and deposition computational models, such as the ones mentioned and reviewed by Tiwari and Kumar [67]. Hence, necessitating detailed fluid dynamics analysis (e.g. Computational Fluid Dynamics—CFD), or the use of in-situ measurements [68].

1.5.1.3 Space Use Attributes

Lastly, the type of activities or the destinated use of a space could also determine the physical vulnerability of a space. Either by limiting the type of materials (e.g., architectural value) or their properties of materials utilized for a certain use (e.g., prevalently asphalt or concrete for driveways), by adding a particular anthropogenic heat source or surface temperature variations (e.g., machinery or traffic), and/or having heavy air pollutant sources. In fact, traffic roads have a maximum albedo or reflectance value to limit the probability of drivers suffering from glare that could result in more traffic accidents [69]. Likewise, governments have also restricted the maximum façade reflectance for buildings next to traffic and rail roads to avoid glare for drivers and other radiation re-direction issues [69]. Chapman and Thornes found that depending on the lane traffic intensity, surface temperatures can suffer systematic variations of approximately 1.5 °C [70].

Differently, Pigeon et al. estimated that in downtown Toulouse (FRA) from February 2004 to March 2005, there was an anthropogenic heat release of around 70 W/m^2 during winter and 15 W/m^2 during summer, while sub-urban areas reported around 1/5th. Highly related to building indoor conditioning during winter, and road traffic during summer [71]. Similarly, Narumi et al. estimated the latent and sensible heat distribution in the Keihanshin district (JPN) from the building's energy performance and road traffic emissions. Thus, displaying higher heat and air pollutant emissions in the downtown, compared to the outskirts (×20 on a total daily) [72].

References

1. UN-DESA (2019) World urbanization prospects: the 2018 revision. New York
2. UN-Habitat (2022) World cities report envisaging the future of cities. ISBN: 978-92-1-132894-3
3. World Health Organization (WHO) (2014) Quantitative risk assessment of the effects of climate change on selected causes of death, 2030s and 2050s. WHO Press, Geneva
4. World Health Organization (WHO) (2016) Ambient air pollution: a global assessment of exposure and burden of disease. World Health Organization
5. United Nations (2015) Transforming our world: the 2030 agenda for sustainable development
6. Gunn SWA (1989) Multilingual dictionary of disaster medicine and international relief
7. Noji EK (1997) The nature of disaster: general characteristics and public health effects. Oxford University Press, Oxford
8. Jamei E et al (2016) Review on the impact of urban geometry and pedestrian level greening on outdoor thermal comfort. Renew Sustain Energy Rev 54:1002–1017. https://doi.org/10.1016/j.rser.2015.10.104
9. Colaninno N, Morello E (2019) Modelling the impact of green solutions upon the urban heat island phenomenon by means of satellite data. J Phys Conf Ser 1343:012010. https://doi.org/10.1088/1742-6596/1343/1/012010
10. Paolini R et al (2014) Assessment of thermal stress in a street canyon in pedestrian area with or without canopy shading. Energy Procedia 48:1570–1575. https://doi.org/10.1016/j.egypro.2014.02.177
11. Stewart ID, Oke TR (2012) Local climate zones for urban temperature studies. Bull Am Meteorol Soc 93:1879–1900. https://doi.org/10.1175/BAMS-D-11-00019.1
12. Fuladlu K (2018) The effect of rapid urbanization on the physical modification of urban area, pp 1–9. https://doi.org/10.14621/tna
13. Salvalai G (2021) Rischio dell'ambiente costruito e dei suoi utenti negli SLow Onset Disasters: fattori tipologici di vulnerabilità ed esposizione negli spazi aperti urbani italiani. In: Design and construction tradition and innovation in the practice of architecture (Progetto e Costruzione Tradizione ed innovazione nella pratica dell'architettura). Enrico Sicignano, pp 1446–1462
14. Intergovernmental Panel on Climate Change (2023) Climate change 2021—the physical science basis. Cambridge University Press
15. UNFCCC (2012) Slow onset events
16. Poli T (2020) Sensible networked façade unit for a healthy and comfortable environment. In: Colloqui AT (ed) New horizons for sustainable architecture, pp 1643–1653
17. European Environment Agency (2019) Air quality in Europe—2019 report. EEA report no 10/2019

References

18. Musco F. Pianificazione urbanistica e clima urbano Manuale per la riduzione dei fenomeni di isola di calore urbano
19. EPA (2008) Urban heat island basics. In: Reducing urban heat islands: compendium of strategies. Heat Island Effect | US EPA
20. Yang L et al (2016) Research on urban heat-island effect. Procedia Eng 169:11–18. https://doi.org/10.1016/j.proeng.2016.10.002
21. Tomlinson CJ et al (2011) Remote sensing land surface temperature for meteorology and climatology: a review. Meteorol Appl 18:296–306. https://doi.org/10.1002/met.287
22. IPCC (2023) AR6 synthesis report: climate change 2023
23. Dyurgerov MB, Meier MF (2000) Twentieth century climate change: evidence from small glaciers. Proc Natl Acad Sci 97:1406–1411. https://doi.org/10.1073/pnas.97.4.1406
24. Lee JY, Kim H (2016) Projection of future temperature-related mortality due to climate and demographic changes. Environ Int 94:489–494. https://doi.org/10.1016/j.envint.2016.06.007
25. CEM-DAT (2023) EM-DAT: the OFDA/CRED international disaster database. Centre for Research on the Epidemiology of Disasters, Universidad Católica de Lovaina, Bruselas
26. Blazejczyk K et al (2012) Comparison of UTCI to selected thermal indices. Int J Biometeorol 56:515–535. https://doi.org/10.1007/s00484-011-0453-2
27. NOOA (2023) Heat index. https://www.noaa.gov/jetstream/global/heat-index. Accessed 23 Mar 2023
28. ISO Standard (2001) ISO 7726 ergonomics of the thermal environment—instruments for measuring physical quantities. ISO Standard 1–62
29. Coccolo S et al (2016) Outdoor human comfort and thermal stress: a comprehensive review on models and standards. Urban Clim 18:33–57. https://doi.org/10.1016/j.uclim.2016.08.004
30. Bröde P (2009) Calculating UTCI equivalent temperature. In: Environmental ergonomics XIII. University of Wollongong, Wollongong, pp 49–53
31. Zare S et al (2018) Comparing universal thermal climate index (UTCI) with selected thermal indices/environmental parameters during 12 months of the year. Weather Clim Extrem 19:49–57. https://doi.org/10.1016/j.wace.2018.01.004
32. British Standard Institution (2017) BS EN ISO 7243:2017—ergonomics of the thermal environment—assessment of heat stress using the WBGT (wet bulb globe temperature) index, 18
33. Mirzabeigi S et al (2021) Tailored WBGT as a heat stress index to assess the direct solar radiation effect on indoor thermal comfort. Energy Build 242:110974. https://doi.org/10.1016/j.enbuild.2021.110974
34. Bernard TE, Barrow CA (2013) Empirical approach to outdoor WBGT from meteorological data and performance of two different instrument designs. Ind Health 51:79–85. https://doi.org/10.2486/indhealth.2012-0160
35. Blanco Cadena JD et al (2021) A new approach to assess the built environment risk under the conjunct effect of critical slow onset disasters: a case study in Milan, Italy. Appl Sci 11:1186. https://doi.org/10.3390/app11031186
36. IEA Buildings. A source of enormous untapped efficiency potential
37. WGBC. Air quality in the built environment—causes of air pollution from the built environment
38. Schneider SH (1989) The greenhouse effect: science and policy. Science (80–) 243:771–781. https://doi.org/10.1126/science.243.4892.771
39. WHO (2016) Ambient air pollution: a global assessment of exposure and burden of disease
40. Niemenmaa V (2018) Air pollution: our health still insufficiently protected. Luxembourg
41. Mintz D (2006) Guidelines for the reporting of daily air quality—air quality index (AQI). United States Environmental Protection Agency, Washington
42. WHO (2005) WHO air quality guidelines for particulate matter, ozone, nitrogen dioxide and sulfur dioxide
43. Cao R et al (2021) The construction of the air quality health index (AQHI) and a validity comparison based on three different methods. Environ Res 197:110987. https://doi.org/10.1016/j.envres.2021.110987

44. Blanco Cadena JD et al (2023) Determining behavioural-based risk to SLODs of urban public open spaces: key performance indicators definition and application on established built environment typological scenarios. Sustain Cities Soc 95:104580. https://doi.org/10.1016/j.scs.2023.104580
45. World Health Organization (2021) Ambient (outdoor) air pollution. https://www.who.int/news-room/fact-sheets/detail/ambient-(outdoor)-air-quality-and-health. Accessed 25 Oct 2021
46. WHO (2018) COP24 special report: health & climate change
47. Meehl GA, Tebaldi C (2004) More intense, more frequent, and longer lasting heat waves in the 21st century. Science (80–) 305:994–997
48. van Vuuren DP et al (2011) The representative concentration pathways: an overview. Clim Change 109:5–31. https://doi.org/10.1007/s10584-011-0148-z
49. Salvalai G et al (2020) Built environment and human behavior boosting slow-onset disaster risk. In: Amoeda R, Lira S, Pinheiro C (eds) Heritage 2020. Coimbra, Portugal, pp 199–209
50. Huttner S (2012) Further development and application of the 3D microclimate simulation ENVI-met. Universitätsbibliothek Mainz
51. Takebayashi H, Moriyama M (2020) Adaptation measures for urban heat islands. Academic Press
52. Abhijith KV et al (2017) Air pollution abatement performances of green infrastructure in open road and built-up street canyon environments—a review. Atmos Environ 162:71–86. https://doi.org/10.1016/j.atmosenv.2017.05.014
53. Kukkonen J et al (2005) Analysis and evaluation of selected local-scale PM air pollution episodes in four European cities: Helsinki, London, Milan and Oslo. Atmos Environ 39:2759–2773. https://doi.org/10.1016/j.atmosenv.2004.09.090
54. Rampini L et al (2020) Monitoring outdoor air quality using personal device to protect vulnerable people. Sensordevices 2020:103–108
55. Mackey C (2017) Wind, sun, surface temperature, and heat island: critical variables for high-resolution outdoor thermal comfort Payette architects. In: Proceedings of the 15th IBPSA conference. Massachusetts Institute of Technology, University of Pennsylvania, United States of America, pp 985–993
56. Lee J et al (2022) A machine learning and computer vision study of the environmental characteristics of streetscapes that affect pedestrian satisfaction. Sustainability 14:5730
57. Jabbari M et al (2021) Accessibility and connectivity criteria for assessing walkability: an application in Qazvin, Iran. Sustainability 13:3648
58. Yang J et al (2020) Urban form and air pollution disperse: key indexes and mitigation strategies. Sustain Cities Soc 57:101955. https://doi.org/10.1016/j.scs.2019.101955
59. Pigliautile I et al (2020) Investigation of CO_2 variation and mapping through wearable sensing techniques for measuring pedestrians' exposure in urban areas. Sustainability 12:3936
60. Whitehouse A (2018) The toxic school run. UK children at daily risk from air pollution
61. Miao C et al (2020) How the morphology of urban street canyons affects suspended particulate matter concentration at the pedestrian level: an in-situ investigation. Sustain Cities Soc 55:102042
62. Tomson M et al (2021) Green infrastructure for air quality improvement in street canyons. Environ Int 146:106288
63. Oke TR (1988) The urban energy balance. Prog Phys Geogr 12:471–508
64. Arnfield AJ (2003) Two decades of urban climate research: a review of turbulence, exchanges of energy and water, and the urban heat island. Int J Climatol. https://doi.org/10.1002/joc.859
65. Chew LW et al (2018) Flows across high aspect ratio street canyons: Reynolds number independence revisited. Environ Fluid Mech 18:1275–1291. https://doi.org/10.1007/s10652-018-9601-0
66. Salvalai G et al (2023) Greenery as a mitigation strategy to urban heat and air pollution: a comparative simulation-based study in a densely built environment. Riv Tema 09. https://doi.org/10.30682/tema090003
67. Tiwari A, Kumar P (2020) Integrated dispersion-deposition modelling for air pollutant reduction via green infrastructure at an urban scale. Sci Total Environ 723:138078. https://doi.org/10.1016/j.scitotenv.2020.138078

References

68. Abhijith KV, Kumar P (2020) Quantifying particulate matter reduction and their deposition on the leaves of green infrastructure. Environ Pollut 265:114884. https://doi.org/10.1016/j.envpol.2020.114884
69. Hjorth E (2017) Glare form photovoltaic systems. Developing an assessment method, 49
70. Chapman L, Thornes JE (2005) The influence of traffic on road surface temperatures: implications for thermal mapping studies. Meteorol Appl 380:371–380. https://doi.org/10.1017/S1350482705001957
71. Pigeon G et al (2007) Anthropogenic heat release in an old European agglomeration (Toulouse, France). Int J Climatol 27:1969–1981. https://doi.org/10.1002/joc.1530
72. Narumi D et al (2009) Effects of anthropogenic heat release upon the urban climate in a Japanese megacity. Environ Res 109:421–431. https://doi.org/10.1016/j.envres.2009.02.013

Chapter 2
User's Factors: Vulnerability and Exposure

Abstract Users' factors highly impact the effective risk due to Slow Onset Disasters (SLODs) affecting public open spaces, such as squares and streets. Individual features that influence users' responses to environmental stressors, because of age, health, gender, habits and behaviour, impact the users' vulnerability. Exposure can then amplify these issues, since it defines the number of users that can be effectively affected by the SLOD. Due to the dynamics in public open space use and SLODs characterization, spatiotemporal approaches are then needed to connect these two basic users' factors. Starting from this perspective, this chapter first traces literature-based correlations between individual features, use behaviour, and individual response to the SLOD-altered open spaces. Then, a novel methodology, to quantify the variations of users' vulnerability and exposure, is offered, to support designers in quickly defining input scenarios for risk assessment and mitigation.

Keywords User behaviour · Exposure · User vulnerability · Heatwaves · Air pollution · Built environment · Public open spaces

2.1 Users' Vulnerability and Exposure: Main Definitions

Assessing users' factors altering Slow Onset Disasters (SLODs) risk is essential in urban Built Environment (BE) as the public open spaces (e.g., squares, streets, and urban parks) [1–4]. Here, users mainly move as pedestrians and spend time in view of social tasks while they are directly susceptible to hazard levels. Furthermore, today, although decision-makers are still generally oriented towards urban or territorial planning solutions, the interest in vulnerability and exposure issues and their assessment approaches is increasing, moving towards the necessity to implement mitigation strategies in a systematic manner by referring to the single public open space scale [5–9].[1] In this context, as for other risks, SLODs can further affect public open space users given their vulnerability and exposure [10, 11].

---

[1] https://www.eea.europa.eu/highlights/protect-vulnerable-citizens (last access: 09/03/2023).

Vulnerability issues concern the "conditions determined by physical, social, economic and environmental factors or processes which increase the susceptibility of an individual, a community, assets or systems to the impacts of hazards" [11]. Concerning SLODs, individual vulnerability is one of the key factors since it alters the users' response to environment stressors. Such differences can be due to age, health status, gender, habits, behaviours and activities performed in public open spaces [1]. In this sense, behavioural issues can alter the possibility to suffer from the emergency effects, as for other kinds of disasters [12] (e.g., by remaining in critical conditions or not [13]).

The impact of SLODs also varies depending on the exposure of users in public open spaces, which could be generally defined as a function of how many users are present in the public open spaces and that could be susceptible to SLOD itself [11].

Furthermore, outdoor and indoor BE functions, urban dynamic patterns and distribution of population flows and densities affect the quantity and typology of users can change over space and time, moving towards the necessity of a spatiotemporal assessment of users' vulnerability and exposure [14–16].

Hence, this chapter summarizes the main vulnerability factors affecting users' SLOD risks in public open spaces, in terms of individual features (Sect. 2.2) and behaviours (Sect. 2.3). And combines them with exposure issues to set-up and propose a methodology that enables to systematically collect and manage users' factors dynamics and variations on a single public open space, adopting a spatiotemporal analysis approach (Sect. 2.4).

2.2 Individual Features

Figure 2.1 summarizes the main individual features affecting user's vulnerability to the intensification of air pollution and heatwaves in public open spaces, as relevant SLODs for the considered urban scenarios, and their connections in terms of effects [1, 6, 8, 9, 14, 17–23].

From a general point of view, all these vulnerability features may alter users' susceptibility to environmental risk, which "describes the predisposition of elements at risk to suffer harm" [24]. Thus, they essentially express the sensitivity to an event potentially harmful to users' health, and so to the potential effects due to SLODs microclimatic conditions which depend on the BE physical vulnerability and from the considered hazard. At the same time, such features impact issues related to: (1) exposure issues (i.e., when, and how users move in the public open spaces and attend them—task performed during their stay); (2) risk awareness and mitigation at the individual level. Moreover, these individual features are interconnected together, as shown by links between them in the left part of Fig. 2.1, as well as effects are correlated too, as shown by the main links in the right part of Fig. 2.1. The schematization in Fig. 2.1 tries to summarize the main correlations between the basic and recognized vulnerability factors. Nevertheless, it is worth noticing that recent works also introduced the "exposome" concept to consider all the exposures a user is subjected

2.2 Individual Features

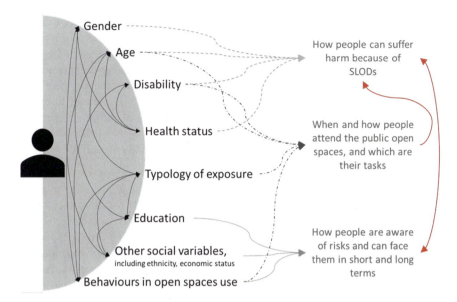

Fig. 2.1 Main individual features affecting social vulnerability to air pollution and heatwaves in public open spaces: interconnection between features (on the left), connections between features and effects (arrows in the central part of the scheme) and interconnection between effects (on the right)

simultaneously to, considering a specific life stage or during the whole lifetime, since the combination of those exposures can differently affect health outcomes for each of the BE users [25].

The combination between age, health status and *gender* are the main features affecting sensitivity towards the risk source, and so the ability of users' bodies "to assimilate, balance and recover from the effects of increasing temperature and air pollution" [14]. In particular, *health status* is one of the crucial aspects, since health fragilities can put users in greater danger when they are exposed to certain hazards. Considering intense air pollution and heat waves, pre-existing diseases such as respiratory (e.g. asthma and allergies), cardiovascular and metabolic (diabetes) diseases and syndromes, especially chronic ones, are the most significant fragilities to be considered as input individual features, as well as to be affected by SLODs conditions [9, 14, 22, 26, 27]. In this sense, they are part of individual sensitivity [18]. Differences due to *gender* impact such results [26, 28], also in view of discrepancy between male and female wages and family care responsibilities, which can alter proper recovery for females [6]. These factors can then be associated with other specific conditions affecting the use of the BE (and so of the public open spaces) and motion within it, such as *disability* and *typology of exposure*.

Disability can alter the possibility to move into public open spaces, and so also the paths selection and the time during which the users are susceptible to the SLODs hazard [29–33]. Moreover, *disability* can alter the users' adaptive capacity and their ability "to access and respond to an early warning system, weather reports, or media for such events" that are correlated with SLODs risks [22]. This last aspect can be amplified by *other social variables*, including ethnicity (i.e. because of language proficiency) and economic status, as well as *education* [6]. These two features are strictly linked, because users with higher education levels could generally gain better economic status, having more resources to face SLODs in short and long terms.

The *typology of exposure* defines which are the habits of the users in respect of urban dynamics and spatiotemporal patterns [1, 6, 16, 34]. Relating to a single public open space scale, typologies of exposure can be quickly organized by mainly distinguishing between:

- *Indoor users, that are residents and non-residents*: they can move in the public open spaces for moving towards/from indoor spaces, and for other social tasks (but with a lower impact in terms of timing). Residents have more prolonged exposure to the general microclimate conditions, but timing issues depend on age, gender, disability, and other social variables (including being occupied or not). In respect to their exposure timing, non-residents include (1) regular visitors/ workers/commuters, as daily or regularly exposed users; (2) tourists and guests of accommodations; (3) one-time/non-recurring users, including customers of buildings open to the public.
- *Outdoor users*: they can be only outdoor users, as passersby, and prevalent outdoor users (users spending time outdoors for leisure/social activities e.g., in cafeteria and restaurant dehors). This includes tourists and visitors. These user type is more susceptible to direct SLODs than indoor users, in view of their position in the BE, although their exposure time is generally lower.

The *typologies of exposure* are also correlated to walkability levels of each public open space as well as of the neighbourhood in which the open space is placed, thus depending on physical factors about the BE and other specific factors that could attract users and promote walking behaviours [1, 20, 35–37]. For instance, an increase in individual walkability at the neighbourhood level could be supported by the adoption of greenery [5, 38]. This solution can also positively contribute to air pollution and heat stress mitigation, and then can decrease exposure to specific pollutants due to vehicular traffic.

Some typologies of users can be more vulnerable than others because of the combination of these individual features. However, some typological conditions could be hence organized also in view of a quick application of assessment tools in real-world BEs by using census data and depending on indoor and outdoor intended uses [34]. In this sense, users' age classification seems to be quick and powerful in defining typological conditions depending on the merged effects of the aforementioned features. This classification can also lead to a strong connection with exposure variations over time (i.e. daily, weekly, monthly) [34]. Main age classes can be hence arranged according to the following age ranges [9, 14, 19, 21–23, 25]:

- *Toddlers (< 5 years-old)*: they depend on their parents/guardians while attending the built environment, thus being not autonomous users. Considering their

2.2 Individual Features

typology of exposure, their presence essentially depends on residential tasks or on the presence of adults that are with them. From a *health status* perspective, they are characterized by physiologic immaturity, developmental changes in body organs and immune system (not yet accustomed to particular critical environmental conditions, especially in case of air pollutants), and the possibility of worsening pediatric diseases in case of acute exposure to high temperatures and pollutants.

- *Young people (5–18 years-old)*: they can be also distinguished depending on the parents' support in the BE use, in terms of parent-assisted children (5–14 years-old) and young autonomous users (15–18 years-old). Considering their *typology of exposure*, their presence in the BE essentially depends on educational, recreative and residential tasks. From a *health status* perspective, SLODs effects are generally lower than those observed for toddlers, although prolonged or chronic exposure to SLODs could generate permanent damage. Furthermore, acute exposure is still critical, since they generally increase emergency department visits (i.e., for asthma or wheezing) for adults.
- *Adult users AU (18–65 years-old)*: they represent the widest class in terms of *typology of exposure*, representing most of the non-residents (i.e., workers and commuters) along with young people attending schools. They are characterized by a larger number of possible *health status* and fragilities, also depending on their consolidated lifestyle (e.g., smokers, low/high physical activity).
- *Elderly EU (> 65 years-old)*: this age class comprehends different abilities in BE use, up to possible *disability* conditions (e.g., in motion, sensory), although they can be characterized by different *typologies of exposure*, a significant part of the can be considered within the residents group, e.g. in view of retirement. *Health status* features essentially depend on different factors, that are also affected by the style of life they have sustained. In this sense, their coping capacity is affected by physical fragility, comorbidity, and less appropriate immune responses. Most of them could have a reduced adaptation to stressful environmental conditions and prolonged exposure to certain hazards could increase the risk of mortality. In particular, they are generally more vulnerable because of the decrease in body temperature regulation because of ageing, thus being particularly susceptible and vulnerable to heat waves.

Furthermore, it is worth noticing that risks, and thus mortality and morbidity, change over the lifetime, not only for respiratory diseases [27] but also for other health issues (e.g. overweight and obesity) [39]. Finally, individual features include also *behaviours* in open space uses. These behaviours can change over the lifetime and they are affected by subjective feedback and objective influencing factors [1]. Besides being correlated with the features shown in Fig. 2.1, subjective feedback is mainly centred on health status, as discussed above, but also depends on users' feelings. According to previous works [1], essential feelings are: (1) driving and needs, including social ones, thus linked to social/physical activities, community life, and social network; (2) physiological feelings, being mainly correlated with climate and environmental factors, and thus also with thermal comfort, with varies over age, gender and individual background and experience; (3) psychological feelings, which are strictly connected to the users' perception of the public open space quality in terms of aesthetics, safety, connectivity/walkability-related aspects and emotional affection. Objective influencing factors, on the other side, are represented by all the in-situ physical open space features (in terms of climate and environmental stimuli, layout, facilities, greenery, intended uses, socialization areas, surface typologies, building components and street furniture, and so on) and background factors referring to the urban BE where the open space is placed (in terms of neighborhood context, district density, location in the urban layout and correlation with the surroundings). As a result, the combination of such *behaviours* with the previous aspects, influence in primis when and how people attend public open spaces, and which are the tasks performed in them. Second, it impacts the way users can face SLODs risks by properly activating prevention, protection and mitigation strategies in both short (immediate response to SLOD stressors) and long (such as lifestyle, possibility to access healthcare facilities) terms. Both these effects alter the level of harm suffered by the user because of the SLODs, as remarked in Fig. 2.1.

2.3 Behavioural Issues in SLODs in Public Open Spaces

Users' behavioural issues in SLODs in public open spaces can be critically distinguished into three topics [1, 13, 35, 40–42], which are the pure behavioural aspects in public open space use, the impact of SLODs on the users' well-being and health, and the correlation with dynamics at the whole urban scale level. These three topics should be discussed by mainly focusing on pedestrians [14], because they are: (1) the fundamental users of public open spaces in urban areas; and (2) often exposed to increasing temperatures (especially during heatwaves) and air pollution due to their movement and use of these spaces. The first topic is represented by pure behavioural aspects, related to the movement and use issues in public open spaces and their surroundings, which essentially depend on the environmental conditions, on the BE use, and on the individual objectives and features summarized in Sect. 2.2. The second concerns the impact that these SLODs can have directly on the well-being and health of the inhabitants. Nowadays, simulation models can be valid tools to understand the probable expansion of the SLODs and certainly to improve and accelerate the comprehension of pedestrian exposure. Those surveys overpass the infield measurements through weather stations of the current conditions through the prediction of their trend thanks to rigorous computational models [43]. Inquiring about the effects on human health and well-being means also understanding the different forms of risk perception and discomfort and the sensations experienced by users, especially in function of the user's typology (i.e., age, gender, and past diseases). In this way, the effects due to "urban stressors" that afflict the well-being of citizens can also be also objectively assessed through the definition of key performance indicators (KPIs) [44]. Finally, also for the impact on the users' health, research has already been undertaken to

2.3 Behavioural Issues in SLODs in Public Open Spaces

determine possible solutions or alternatives to reduce exposure thanks to methodologies that re-design and mitigate the influence that the use of specific building components, materials, layout, urban furniture, nature-oriented solutions [45, 46]. The last topic serves to direct the attention no longer on the behaviour or health of the single user but on the community as a whole, as well as the correlation with dynamics at the whole urban scale level [16, 40, 47]. Indeed, analysis models and implemented solutions should be based on key findings retrieved at a more detailed scale, e.g. relating to single public open spaces.

Nevertheless, the basic step for the second and third topics is represented by an organized overview of the users' behaviours. This section focuses on the first topic. Recent review work defined a systematic classification of users' behaviours in public open spaces to a general-to-specific scale approach [1], by essentially distinguishing (1) specific movement (walking, sitting, thermal adapted behaviours), (2) categories of behaviours with the same attributes (i.e. physical activity, leisure activity, social activity), and (3) general occurrence behaviours, that are related to "site attendance without activity type specification". Figure 2.2 adopts and expands the concept of this previous classification, moving towards the definition of dependencies among blocks, and introducing simulation tasks about use and movement in public open spaces. Figure 2.2 also connects behavioural issues with users' typologies, in terms of built environment use, individual features (as in Sect. 2.2) and the role of "attractors and repulsors" within the built environment. From a general point of view, the users' movement can be divided into path selection, trajectories and speed issues. Then, use issues are herein considered in their essential connection with the permanence of users in every single open space, aimed at accomplishing the individual objectives and purposes in it.

2.3.1 Specific Movement and Behaviours

According to the adopted base classification [1], users' movement can be connected to walking behaviour, as well as to thermal adaptation behaviours, while space use can be related to staying/sitting behaviours, and to thermal adaptation behaviours. However, behaviour attributes affect the specific movement adopted by each user, and they can be essentially distinguished into:

- Physical activities, defined as all the activities that imply energy expenditures for users' leisure, transport, work purposes [1]. In this sense, these activities highly impact physiological implications on the users, varying the metabolic rate, and thus the internal heat production as well as the amplification of heat stress [48]. As a consequence, they can strongly affect thermal adaptation, since they modify the human core temperature [49]. Similarly, they can also alter individual safety, especially in case of very hot and extreme heat conditions that can be considered in a SLOD scenario [50, 51].

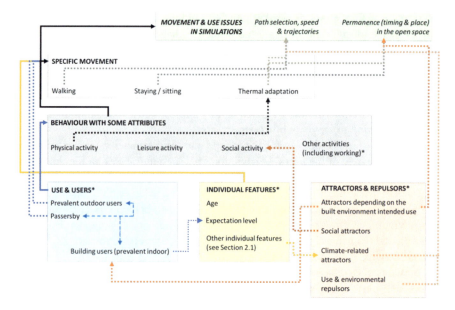

Fig. 2.2 Users' behaviours in SLODs in public open spaces, by pointing out connections between the use and movement simulation tasks with specific behaviours, behaviours with attributes, typologies of users depending on the built environment use, individual features (compare with Sect. 2.1) and attractors and repulsors in the built environment. Main connections between blocks are shown by continuous lines, while dependencies between specific elements of different blocks o within the same block are respectively traced by dotted lines and dashed lines. The previous classification from [1] has been integrated by adding blocks and elements marked by *

- Leisure activities, such as sitting, relaxing, strolling, visiting, sightseeing, or performing other enjoyable activities during free time, including recreation and physical activities [1]. In general terms, they are connected to positive emotions for the users, and they also depend on the users' age and habits, being associated with social activities, especially for the elderly.
- Social activities, mainly related to the interactions between the users in outdoor, for different purposes [1], including physical and leisure activities, as well as age and gender. Some works also pointed out that social activity can increase individual thermal adaptation, because it positively affects the users' psychological adaptation promoting the presence of users in open spaces [52].
- Other activities (including working), which are introduced by this work to manage additional users' objectives connected with the public open space fruition. They can significantly vary the time extent and the conditions under which the users remain in the open spaces, which can be obliged and not spontaneous. Thus, different effects of SLODs on the users can be noticed, especially in terms of the risk to their health, because outdoor workers are generally more vulnerable to air pollution than indoor workers [51, 53].

2.3 Behavioural Issues in SLODs in Public Open Spaces

Literature is focused on the user's thermal adaptation as one of the main behavioural drivers [1]. Besides the effects due to behaviour attributes (i.e. in physical activities and in connection with psychological adaptation), the thermal adaptation depends on different aspects combining the outdoor microclimate conditions with the individual features, i.e. individual age, clothing, expectation level in outdoor fruition, fruition time in outdoor (i.e. short term, up to 15 min, or longer-term, such as in 1 h behaviours) [13, 49, 54]. From a general standpoint, adaptation concepts are based mainly on thermal sensation votes, assuming that users will prefer selecting conditions with neutral temperatures. Nevertheless, considering that thermal comfort is a multidimensional concept, the alliesthesia concept has been also included in literature, to point out that "the pleasure or displeasure of a sensation is not stimulus bound but depends on internal signals" [55]. In this sense, recent works underlined that "Adaptation predicts seasonal differences in dimension of thermal sensation", while "Alliesthesia predicts seasonal differences in dimension of thermal pleasure" [56], thus underlining how differences due to the seasonal context of the built environment exist and they could then influence the occurrence of extreme events also related to SLODs.

2.3.2 The Role of the Built Environment: Use and Related Users' Typologies, Attractors and Repulsors

Considering the intended use of the built environment and their users' typologies, three different categories of pedestrians can be noticed [34]. Passersby are assumed as the users that do not remain within the considered open space for a long time, e.g. just to cross it and move towards another part of the urban fabric, or walk in it for less than 15 min. Prevalent outdoor users are visitors/tourists, customers of dehors, and other users spending most of their time outdoors for social or leisure activities. Users attending mass gatherings or temporary special events hosted in the public open spaces can be included in this category too. Thus, they are associated with higher exposure times outdoor. Differently, building users mainly attend indoor areas, thus becoming passersby or prevalent outdoor users depending on secondary tasks in the built environment fruition.

These users' typologies interact with the elements composing the public open space and the facing built fronts, which can be associated with their attractive or repulsive effects on them. In this context, Fig. 2.3 provides a scheme of main attractors depending on the built environment use and on social and climate-related aspects, and the main repulsors. Therefore, these elements directly modify the users' decision process for path selection, trajectories, and permanence in public open spaces. Simulation models should take into account them to more effectively differentiate users' responses, behaviours and SLOD-related effects due to exposure levels over time. The path selection takes into account phenomena at the whole "urban/district/neighbouring level" and at the "single open space level". At the "district/neighbouring level", the users' paths are composed of movement tasks towards and into several open spaces, that are streets, squares, and other public areas (including parks, porticos and gallerias). At the "single open space levels", the path selection can be overlapped with the users' trajectories in the built environment (thus, both indoors and outdoors). Therefore, at both these levels, permanence issues are related to the time exposure of each user in the given environmental conditions and to the specific position, which is connected to path selection and trajectories. According to previous works [34, 42, 57–65], path selection, trajectories and/or permanence mainly depend on factors which are, regardless of the SLOD conditions:

- individual objectives in motion (i.e., intended travelling reason, length of the walking travel and time, familiarity with the built environment, safety perception).
- starting points and targets for motion (e.g., residential, office and other buildings) attracting users for their use while open to the public. Or special buildings as main users' attractors (e.g., theatres, museums, religious buildings, government buildings, metro/train stations, hospitals, schools, universities), as well as outdoor areas intended for rest or gatherings (e.g., dehors), as in Fig. 2.3.
- attractiveness of the built environment (due to available public amenities and services), urban quality and/or morphology (e.g., slope, streetscapes, type of road,

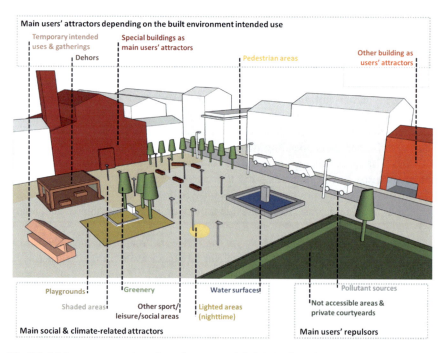

Fig. 2.3 Main attractors and repulsors for users and their behaviours in public open spaces in the urban built environment

2.3 Behavioural Issues in SLODs in Public Open Spaces

quality of pedestrian ways), urban furniture, greenery and others influencing walkability (e.g. surface typologies, presence of architectural barriers to movement, the presence of dedicated pedestrian areas as in Fig. 2.3).
- And, other environmental features that are not directly related to SLODs (e.g., noise, construction or, during nighttime, the lightning level as traced in Fig. 2.3).

As such, the degree of attractiveness of the built environment areas strongly depends on the typologies of users involved in their use, thus also implying daily, weekly, monthly, seasonal and yearly variations [1, 34, 66]. The main differences, especially considering a historic built environment, could be related to residents, non-residents and tourists in the public open spaces. For instance, as shown in Fig. 2.3, tourists' movement and permanence are oriented towards sights and main urban attractors, that are special and public buildings, pedestrian areas, temporary gatherings, leisure/social areas, green areas, public transport stations, other buildings such as facilities (including medical ones) and shops [63]. As a consequence, experience times in outdoor areas can represent one of the most significant in a urban scenario in terms of relevance. However, related temporalities depend on daily visitors and touristic flows [67]. Additional differences in attractors and repulsors in the public open space are, however, due to the users' age typologies [1]. For instance, indeed, regardless of SLOD and climate features, playgrounds and sport areas are essential attractors for toddlers and young people for different leisure and social activities, while the effective attractors for adults and elderly depend on performing social sedentary activities (e.g., areas with benches) or non-sedentary activities (e.g., physical activities such as jogging or walking, in areas with greeneries or large pedestrian areas). Indeed, temporalities in use and permanence depend on their age-related characterization, as discussed in Sect. 2.1 [34].

Furthermore, the presence of other users in the environment introduces additional attraction and repulsion phenomena at both wider and smaller scales, which can be related to (1) group effects among users (e.g. sharing the same objectives and performing behaviours with some attributes together) and (2) pedestrian flows/density (i.e., influence on movement direction according to different motion heuristics, retention of the individual desired speed and of the minimum/desired physical distance from surrounding users and obstacles) [30, 48, 57, 60, 68].

2.3.3 SLODs Conditions Affecting Users' Behaviours: Air Pollution and Increasing Temperatures

Users' behaviours are also affected by the effective combination of such built environment attractors and repulsors with climate/weather conditions, and so with SLODs features. In specific, increasing temperatures (hence, more intense heat waves) and air pollution implying heat and air quality distress on the users [13, 15, 48, 57, 69–74]. SLODs (and thus climate conditions) first affect users' behaviours in terms of immediate/short-term choices, referred to movement and use issues as introduced in

Fig. 2.2. Weather parameters such as sunlight and precipitation directly affect users' behaviours [1, 57, 72, 75]. But, it can be assumed as main driver air temperature, which can be also assessed in terms of thermal perception rating with the universal thermal climate index—UTCI, Physiological equivalent temperature—PET, or other heat stress rating indices [76].

These behaviours can be considered as adaptive behaviours that are combined to the other individual responses in trajectories and heuristics in path selection, providing "adjustments made by the people in their normal outdoors behaviour to heat" [69]. Previous works provided insights on the Neutral PET (NPET), which "refers to the PET value at which people feel neutral (not too cold or too hot)" [77], and which can vary depending on age-related factors in view of the users' physiological issues. In this sense, additional differences in NPET and also preferred PET between residents and tourists have been noticed too, also pointing out that tourists can be characterized by quite larger preferred boundaries in Thermal acceptability ranges than residents [78]. Nevertheless, one of the most powerful approaches to users' movement and permanence in public open spaces is based on the correlation between air temperature or heat stress indices (UTCI and PET), and thermal acceptability [13], to define the probability that a user could spend time in a given part of the built environment depending on its microclimate condition. Research showed that transient thermal acceptability, which can be associated with passersby (exposure time up to 15 min), is generally higher than the 1 h thermal acceptability, which can be associated with prevalent outdoor users, since the formers are essentially involved only in walking behaviours while travelling. Results suggest that such correlations are limitedly affected by age of individuals, thus boosting their application, as shown in Chap. 3.

Nevertheless, it is worth noticing that further individual features could alter the final probability to move and remain in a certain microclimate condition, along with other boundary conditions as discussed in Sect. 2.2. For instance, recent research has pointed out that building users can be characterized by a higher mean expectation level of using outdoor spaces than outdoors users, and that this expectation levels grow along with the increase of indoor-staying time length [54]. Consequently, users "who stay indoors have a lower tolerance of unwanted outdoor thermal conditions than those who are already in the outdoor environment". Nevertheless, these results are meant for intermediate seasons from experimental data, underlining the necessity to perform additional tests to better connect such issues with extreme built environment conditions.

Beside issues concerning path selection and trajectories, temperature also provides significant insights into the users' speed. Although preferred speed is a function of the individual features, and of circumstantial inputs, such as the environment where the user moves, the reasons for movement, and the surrounding pedestrian density [30, 68], previous research generally pointed out that users tend to walk faster when temperatures are outside their comfortable range [69]. In fact, they would like to reduce the time of exposure to uncomfortable scenarios as long as possible. Previous works also investigate the effect of walking speed on heat stress [48]. Stating that heat-stress-optimal walking speed can be reached by estimating its values for a wide

2.3 Behavioural Issues in SLODs in Public Open Spaces

range of air temperatures with the use of computational modelling of metabolic heat production and body thermal regulation. However, in hot urban environments, the increased walking speed would imply that people are producing extra heat amplifying their heat stress.

Although the theoretical Energy-expenditure-optimal walking speed is essentially independent form environment conditions (about 1.21 m/s), the heat-stress-optimal walking speed depends on climate and walking distance. In particular, due to heat stress-thermal regulation phenomena, the walking speed increases with the reduction of the distance, while it moves towards convergence when travelling distances are longer than 500 m, which are compatible with movements between different open spaces. Nevertheless, the city's pace of life modifies the impact on heat stress at the users' level (e.g. experimental speeds are, on average, 1.34 m/s > 1.21 m/s as the theoretical optimal one) [48]. Similarly, some works underlined how other features of the weather and the surroundings affect the pedestrian speed [72], by also including other BE conditions (i.e., the presence of snow on the ground, the height of buildings in the surrounding that constitutes shading and protection from winds) as additional stressors.

Furthermore, increasing temperature and air pollution (i.e. particulate matter) have a specific influence not only on the individual motion in outdoor, but also on the pedestrian volume as a whole [70, 71, 75]. Such conditions alter the basic pedestrian volume, which can be reasonably related to density, diversity of land use, and design including safety and amenity, and destination accessibility and distance to transit [62], as discussed in Sect. 2.2.2. The assessment of SLOD-affecting pedestrian volume becomes essential to determine the exposure for pedestrians, and, mainly, considering passersby [14]. Previous research investigated regressions among weather parameters and the related walking volume by dethatching it from video analysis from Web cameras pointed in specific locations of the urban public open spaces [75]. Regression results suggest that PM10 concentration determines users' intention to walk thus the pedestrian volume [71]. Experimental analysis in real-world scenarios confirmed that exposure time thresholds to particulate matter conditions can push pedestrians to avoid open spaces and outdoor activities while facing critical (and short-time) environmental conditions, and to refuge in indoor public buildings such as commercial spaces with free entrance [15]. Similarly, considering indoor areas and warm climate conditions, users in public open spaces also tend to deviate their routes and trajectories making choices in relation to the possibility to take a break inside public/shopping buildings with cooler air temperatures.

In this context, additional implications of SLODs are noticed not only in peak walking volume differences, but also in long-term choices, thus concerning pedestrian habits (e.g., starting to change travel modes, avoiding taking the shortest and busiest path by foot but preferring a longer street with a low rate of air pollution and perhaps crossing a green area of the city) [58]. Experimental models defined the choice probability between moving on feet or with private vehicles, discussing individual exposure to air pollution, especially to PM10 and PM2.5 [70, 71]. At the same time, users who chose to cover certain paths walking rather than driving their car can be excessively exposed to such air pollution, and the additional travel time involved in walking increases sensibly pedestrian exposure to particulate matters. Moreover, policies that encourage to walk instead of driving could be harming pedestrian's wellbeing; strengthening the importance of separate routes for pedestrian and vehicle traffic.

In view of the above, Table 2.1 tries to trace a general summary of how the citizens interact with the built environment in discomfort conditions due to increasing temperature and air pollution, by tracing a short qualitative description of probable behaviours and the outcoming main built environment implications. Hence, pedestrians generally prefer:

- shaded area and, mainly, urban canyons in summer conditions, either by trees or the self-shading from the buildings, thus looking for protection from direct solar radiation;
- low traffic level street corridors, since trafficked roads can be perceived as one of the basic repulsors being pollution sources;
- areas with greater greenery or water bodies coverage;
- canyons where the wind flow is favoured in summer;
- surfaces with low surface temperatures in summer.

Table 2.1 Main probable users' behaviours in SLODs scenarios in public open spaces, and their implications on the built environment

SLOD scenario	Main probable users' behaviors	Main implications on the built environment
Warm—non polluted	Users tend to move towards more shaded pathway/corridors/areas. They can prefer (a) to be near green and water bodies, and, often, (b) to access air-conditioned services (e.g. shops, restaurants) which provide cooler air temperatures. Moreover, lighter and lower surface temperatures superficies are generally preferred	Increased crowd level in outdoor areas or streets, i.e. secondary streets, green corridors, water areas, urban forests, shaded areas. In addition, larger use of automobiles is expected
Warm—polluted	Same considerations of the warm-non polluted scenario prevail. In addition, users tend to move through and closer to less crowded pathways/corridors/areas, by avoiding trafficked street corridors when possible	Increased crowd level in outdoor areas or streets, i.e. secondary streets. In addition, larger use of automobiles and mid-traffic levels in every street type are expected
Mild/cold—polluted and mild/cold—non polluted	In this case, users will tend to move through and closer to less shaded areas (i.e. sunny) and, even if polluted, crowded pathways/corridors/areas are preferred. Moreover, trafficked street corridors could represent a source of heat, becoming more attractive. In addition, users can prefer remaining indoors and transferring in automobiles	Decreased crowd level in outdoor areas or streets, since users tend to remain indoors. Higher automobile use, favoring traffic and pollution production

These favourable built environment configurations against SLODs embody an attractive factor, especially for those categories of people with frail health, elders, or youngsters constituting for them a source of well-being. At the same time, if the delineated probable behaviours are performed by the majority of the public open spaces users, they inevitably lead to criticalities that have to be considered (e.g., overcrowding along the preferred paths, large use of private automobiles with the consequent increase of particulate matter emissions).

Finally, pedestrians could be also guided towards the adoption of specific resilience-increasing behaviours, by actively informing them and supporting their routing for healthy, wellbeing-oriented and safe choices in movement and open space use [58, 79–84].[2] Guidance solutions are essentially based on the communication of alternative routes to each of the pedestrians, using mobile applications, and adapting navigation algorithms according to local data (e.g., real-time monitored ones). Thus, this kind of guidance is oriented towards the increase of quality awareness for pedestrians [80], and they can reach safety improvement as a consequence of alternative behaviours from the aforementioned ones aimed at individual adaptation because of built-environment features, climate conditions and implemented mitigation solutions in the public open spaces (e.g. presence of shading and cooled areas, trees, separate routes for vehicles and pedestrians).

Different approaches can be defined, by mainly applying them to the path selections at the whole "urban/district/neighbouring level" rather than the microscale. In fact, pedestrians have to be mainly informed about the next areas during their journey, because they are not aware of non-visible and non-directly experienced conditions, rather than of surrounding scenarios where they can perform the aforementioned adaptive behaviours. Pedestrian paths can be planned depending on each SLOD in a separate manner, e.g. in terms of pollution sources affecting air quality [81, 82], or heat stress risk [79]. In previous research [83], the routes calculated depending on the improvement in the levels of low air pollution inhalation (by esteeming the inhaled mass of fine particles—PM2.5) were compared against the traditional shortest paths. Results demonstrate that traditional pedestrian trips are shorter than healthy walking ones, but they show significant benefits for human health. Routing could also combine the attractiveness of urban environment elements while protecting pedestrians from exposure to air pollution and also high noise levels, suggesting choosing more illuminated paths and proximity of a green area, shaded paths and the avoidance of hottest urban canyons [58]. Approaches could be specialized towards pedestrians and cyclists [82], while multimodal routing frameworks have been proposed [84], by allowing users to select and compare different approaches (fastest, shortest, least exposure to air pollution, balanced) by different means of transportation (e.g. car, motorbike, bicycle, on foot).

---

[2] https://eit.europa.eu/news-events/news/ambiciti-first-mobile-app-street-level-air-and-noise-pollution-launches-europe.

2.4 Users' Factors Dynamics

The users' exposure and vulnerability vary over time depending on the built environment use and the urban dynamics, and so the SLODs risks are affected by the spatiotemporal distribution of users [14, 16, 34, 67, 85, 86]. Two recent methodologies were developed within the BE S^{2}ECURe project to take into account how the factors described in Sects. 2.1 and 2.2 could be arranged over space and time, to additionally derive typological conditions of users-related factors in public open spaces [14, 34]. In detail, these methods are focused on urban squares as application contexts, although they can be replicated for streets, as well as for small districts within the urban areas. Furthermore, these methods rely on the use of a remote analysis approach ensuring a quick quantification of how many users are exposed over time, depending on their position in the built environment, and their main features. Such an approach could boost the application by local authorities and designers, who can provide a preliminary but reliable overview of assessed factors and then can deepen the level of knowledge of phenomena using detailed in-situ surveys and monitoring activities [57, 66, 72, 75, 87]. Figure 2.4 shows the overall assessment methodological framework, which is composed of the following steps:

1. Detection of indoor and outdoor intended uses;
2. Association of quick occupant loads for each detected intended use, by associating main exposure temporalities;
3. Assessment of typologies of users in terms of individual vulnerabilities, and temporal variations (e.g., also due to timetabling and users' habits);
4. Time-dependent assessment of users' exposure (number of people) and social vulnerability (by typologies of users) for each intended use and for the whole public open space, thanks to selected KPIs.

The proposed methodological steps are described with reference to a single square, according to Sect. 2.1 and by considering the relevant role of this public open space in the historical urban built environment.

2.4.1 Detection of Indoor and Outdoor Intended Uses

The remote detection of indoor and outdoor intended uses in the built environment can be supported by GIS tool, where available, or even data from web mapping platforms (e.g., Google Maps or Open Street Maps). These data sources allow primarily detecting facilities open to the public that are hosted in indoor and outdoor areas (e.g., office; dehors; religious, cultural or government public buildings; restaurants; shops), and private areas in the built environment (e.g., homes, private courtyards). These outdoor and indoor areas can be grouped into homogeneous categories, respectively according to Tables 2.2 and 2.3, depending on their main typology of users (Sect. 2.2). In particular, for outdoor areas, the main categories are Passersby as

2.4 Users' Factors Dynamics

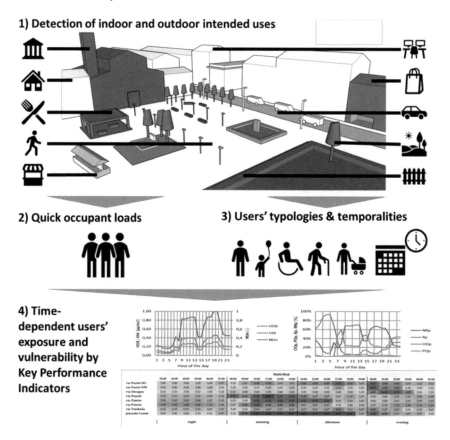

Fig. 2.4 Users' vulnerability and exposure assessment framework derived from the results of the BE S²ECURe project [14, 34]

Only Outdoor Users (OO), and Prevalent Outdoor Users (PO), which are the users of dehors, like open-air terraces of bars and restaurants, and users in gathering areas. For indoor areas, and thus buildings, the main categories are Non-Residents (NR) and Residents (R). In particular, the attractiveness and comfort issues in indoor areas are also included in the classification provided by Table 2.3, and different intended use for each class can be then defined also in view to their connection with specific occupant loads, as discussed in Sect. 2.4.2. The provided assumption is herein based on Italian data and regulations, but it can be considered valid even for other international contexts sharing similarly intended use typologies. For each indoor and outdoor area, the available gross surface of the ith area SU_i (m²) can be estimated through GIS tools. In case no structured data is present, web mapping platforms can be used indeed. For instance, OpenStreetMap[3] is a free, open geographic 2D and

---

[3] https://www.openstreetmap.org/.

Table 2.2 Outdoor areas classification based on main users' typologies and their familiarity with the space

Type of outdoor users (ID)	General assumed familiarity	Classes: main definition
Passersby as only outdoor users (OO)	Unfamiliar	Pedestrian areas, sidewalks, accessible parks, and green areas: users performing walking behaviours
Prevalent outdoor users (PO)	Unfamiliar	Dehors, like open-air terraces of restaurants and bars, open markets, other outdoor areas with specific intended use: users performing walking and staying/sitting behaviours
		Outdoor temporary gathering areas, such as for special outdoor intended uses and mass gatherings: users performing walking and staying/sitting behaviours

3D database updated and maintained by a community of volunteers, Calcmaps[4] is a freeware tool to perform measurements on aerial views, and, Google Street Maps[5] can be used to visually derive the number of floors and manually record them to assess SU_i.

Since the open space is part of a wider urban layout. The following assumptions allow defining the effective perimeter to be assessed, in relation to a given public open space [16, 88]:

A. Considering the analysis of indoors, all the buildings with direct access to the assessed square should be considered, because they can be internal starting points/targets for pedestrian movement in the square. The presence of doors, passages, or gates connecting indoor and outdoor areas can be also retrieved by Google Street View.[6]
B. According to a conservative approach, users, who may perform walking behaviours while entering/exiting the square, can alter exposure and vulnerability within the public open space itself. For this reason, the assessed area should also comprise half the streets linked to the square, due to shortest path principles as a simple heuristic in pedestrian routing. This assumption means that all the indoor and outdoor areas along these streets have to be identified and accounted for.
C. Obstacles and areas not accessible to the users, such as monuments, fountains, fenced areas, and water bodies, can be reasonably excluded from the presence of users (0.00 pp/m²).

---

[4] www.calcmaps.com/it/map-area.
[5] www.google.it/maps/?hl=it.
[6] https://www.google.com/intl/en/streetview/.

2.4 Users' Factors Dynamics

Table 2.3 Indoor areas (buildings) classification based on main users' typologies and their familiarity with the space

Type of indoor users (ID)	General assumed familiarity (secondary assumptions[a])	Classes: main definition	Specific intended uses
Non-residents (NR)	Unfamiliar (familiar)	Sensitive buildings: social or healthcare-related uses with possible high concentration of vulnerable users because of their age, health frailties, economic status	Educational buildings except universities
			Hospitals and healthcare buildings
			Social welfare facilities
	Unfamiliar (familiar)	Commercial buildings, services and other facilities, cultural and religious buildings: buildings open to the public, with potential high number of contemporary users, and possibility of air-conditioning implementation	Shops, bars, restaurants, other commercial buildings
			Universities
			Government buildings, police stations, administrative buildings, front office spaces
	Unfamiliar		Religious buildings, museums, other cultural and historical buildings
			Transport services and stations (e.g., railway stations, metro stations)
	Unfamiliar (familiar)	Business: buildings that can be also open to the public, but with lower potential number of contemporary users (mainly, workers), and possibility of air-conditioning implementation	Office buildings also open to public, including banks, insurance, professional studios, research center, specialized services
			Office and business spaces close to the public
	Unfamiliar (familiar)	Production: private factories generally reserved for authorized users	Factories, couriers, warehouses, construction sites, labs, workshops

(continued)

Table 2.3 (continued)

Type of indoor users (ID)	General assumed familiarity (secondary assumptions[a])	Classes: main definition	Specific intended uses
	Unfamiliar	Accommodations: users are visitors, but they are present for more than one night, also during nighttime	Hotels and other accommodation facilities
Residents (R)	Familiar	Residential: citizens having the most long-lasting exposure	Homes and other residential buildings (including monasteries, student residences)

[a] Users can be considered familiar in case of workers in public spaces are generally very relevant in respect of the visitors

2.4.2 Quick Occupant Load for Users' Exposure and Related Temporalities

Tables 2.4 and 2.5 resume the maximum occupant load OL_i (pp/m^2) and main proposed timetable for outdoor and indoor areas, respectively, derived from the approach of the BE S^{2}ECURe project [34]. The occupant load should be applied to the timetable range, while 0 pp/m^2 should be considered out of the addressed timetable. The maximum occupant load allows to calculate the number of users according to a quick approach and by adopting a conservative standpoint on maximum values. Specific data based on surveys or remote data collection could certainly be used; for example, for hospitals and accommodations: the number of available beds; for restaurants, cinemas and theatres: the number of seats. Working days and holidays should also be distinguished to assess conditions between recurring conditions during the year and Sundays and other national Holidays, as occupation will increase or decrease accordingly. In addition, seasonal variations can be taken into account by modifying the maximum loads, especially for tourist destinations, and considering outdoor areas. In Table 2.4, pedestrian areas are associated with specific scenarios in terms of pedestrian volume, according to average occupant load derived from the levels of service, rounded at the next 0.05 pp/m^2 [89]. In case of no specific data about pedestrian volume, level of service A should be assumed, at least, since this standard value allows pedestrians to move freely. Considering tourist destinations, holiday occupant load should be higher than the one related to working days (e.g., 0.1 pp/m^2 for working days and 0.2 pp/m^2 for holidays). Dehors are assumed according to Italian fire safety occupant loads [90] for the same intended use. In Table 2.5, data are derived according to Italian fire safety occupant loads [90], also considering possible alternatives derived from previous works of the research group [67].

2.4 Users' Factors Dynamics

Table 2.4 Occupant loads and generic timetable for outdoor areas, according to Table 2.2, derived from the approach of the BE S²ECURe project [34]

Classes	Specific scenarios	Occupant load OL_i (pp/m²)	Generic timetable (W = working days; H = holidays) (h)
Pedestrian areas, sidewalks, accessible parks, and green areas	Level of service A	0.10	W and H = 7–24
	Level of service B	0.20	
	Level of service C	0.35	
	Level of service D	0.60	
	Level of service E	1.05	
	Level of service F	> 1.05	
Dehors, as open-air terraces of restaurants and bars, open markets, other outdoor areas with specific intended use		0.4 (except from specific intended uses)	W and H = depending on the activity timetable
Outdoor temporary gathering areas		> 2.0 (up to 4.0)	W and H = depending on the temporary intended use

In view of the above, Tables 2.4 and 2.5 describes only the exposure dynamics, in terms of minimum (closure of activities, thus implying users as not present) and maximum (full opening of activities), but they do not quantify or qualify the users' vulnerability.

2.4.3 Users' Vulnerability and Related Temporalities

Apart from the users' vulnerability related to their position in the built environment and familiarity with it (by dividing Only Outdoor Users, Prevalent Outdoor Users, Non-Residents and Residents and defining their related temporalities as in Tables 2.4 and 2.5), additional individual features that can be assessed in a quick manner are expressed in terms of users' age, gender and general health status. Such data can be collected from local, regional and/or national census and statistics databases [91]. Considering the Italian context, annual reports from ISTAT provide basic statistics in terms of the percentage distribution by population, also detailing data for each

Table 2.5 Occupant loads (based on Italian fire safety codes [90]) and main timetable for buildings, according to Table 2.3, derived from the approach of the BE S²ECURe project [34]

Classes	Specific intended uses	Occupant load OL$_i$ (pp/m^2)	Generic timetable (W = working days; H = holidays) (h)
Sensitive buildings	Educational buildings except universities	0.4—as an alternative, a maximum of 26 individuals in each classroom and annex (e.g., refectory, gym) + 4% for teachers and other workers, rounded to the upper integer	W = 8–14 (expect of other lessons timetabling)
		0.1	W = 14–18 (except of other lessons timetabling)
	Hospitals and healthcare buildings	Ambulatory and similar, 0.1	W = 8–20 (except of other detailed timetabling)
		Wards (inpatients), 0.1 or the number of beds	W and H = 0–24
		Spaces for visitors, 0.4	W = 8–20 (except of other detailed timetabling)
		As an alternative, the number of in-service personnel plus the average number of visitors and inpatients referring to at least three typical days	W and H = depending on the timetable
	Social welfare facilities	Closed to the public, 0.1; open to the public, 0.4; gathering areas open to the public, 0.7	W and H = depending on the timetable
Commercial buildings, services and other facilities, cultural and religious buildings	Shops, other commercial buildings	0.4	W and H = depending on the activity timetable
	Bars, restaurants	0.7	W and H = depending on the activity timetable
	Universities	0.4	W = 8–20 (except of other detailed timetabling)

(continued)

2.4 Users' Factors Dynamics

Table 2.5 (continued)

Classes	Specific intended uses	Occupant load OL$_i$ (pp/m^2)	Generic timetable (W = working days; H = holidays) (h)
	Government buildings, police stations, administrative buildings, front office spaces	Closed to the public, 0.1; open to the public, 0.4; gathering areas open to the public, 0.7—as an alternative, the certified number of occupants (e.g., workers) + 25%, rounded to the upper integer	W and H = depending on the activity timetable
	Religious buildings	0.7—as an alternative, the number of seats plus the number of standing places	W and H = depending on the celebration timetable (suggested only in H)
	Cinemas and theatres	3 (applied to the area open to the public)—as an alternative, for theatres and cinemas, the number of seats for the public + 20% for workers, rounded to the upper integer	W and H = depending on the activity timetable
	Museums, other cultural and historical buildings	0.4 or 0.7—as an alternative, derived from visitors' data provided by tourism organizations	W and H = depending on the activity timetable
	Transport services and stations (railway stations, metro stations)	0.2 (extended to the whole building area)	W and H = 0–24
Business	Office buildings also open to public, including banks, insurance, professional studios, research center, specialized services	Closed to the public, 0.1; open to the public, 0.4; gathering areas open to the public, 0.7—as an alternative, the certified number of occupants (e.g., workers) + 25%, rounded to the upper integer	W and H = depending on the activity timetable

(continued)

Table 2.5 (continued)

Classes	Specific intended uses	Occupant load OL$_i$ (pp/m^2)	Generic timetable (W = working days; H = holidays) (h)
	Office and business spaces close to the public	Closed to the public, 0.1; gathering areas for workers, 0.7—as an alternative, the certified number of workers + 25%, rounded to the upper integer	W and H = depending on the activity timetable
Production	Factories, couriers, warehouses, construction sites, labs, workshops	Closed to the public, 0.1; open to the public, 0.4; gathering areas open to the public, 0.7—as an alternative, the certified number of occupants (e.g., workers) + 25%, rounded to the upper integer	W and H = depending on the activity timetable
Accommodations	Hotels and other accommodation facilities	0.4—as an alternative, the number of beds + 20% for workers, rounded to the upper integer	W and H = 0–24
Residential	Homes and other residential buildings (including monasteries, student residences)	0.05	W and H = 0–24

municipality.[7] Municipalities-related vulnerability distributions can be considered valid for the public open space in the urban fabric, but specific surveys can be then provided at the neighbouring or single open space-level to refine statistics. Table 2.6 resumes data about residents and non-residents in respect of age typologies, in view of the different habits of these users in the built environment fruition [34]. Temporalities are herein also considered by means of a "presence coefficient" cp, which can be equal to: 1, when users are present; 0, when users are absent; 0.09, to consider users spending their time at home since they are unemployed. Nevertheless, ad-hoc assignment of cp could be provided according to specific case study data. For each age range a in Table 2.6, the percentage of users by age AP_a (%) is associated with the aforementioned census-based statistics. To consider gender-related vulnerabilities, the percentage of males Mp and females Fp (%) can be then multiplied by these

---
[7] Age and gender (2020): http://demo.istat.it/popres/index.php?anno=2020&lingua=ita; health: https://www.istat.it/it/archivio/5471 (last access: 25/07/2021—in Italian, English version available at https://www.istat.it/en/).

2.4 Users' Factors Dynamics

Table 2.6 Users' vulnerability according to age typologies, as introduced in Sect. 2.2, and derived from the approach of the BE S²ECURe project [34]

Users' typology depending on age classes (year range, motion features)	Residents: timetable (h); cp = coefficients of presence (notes)	Non-residents: (timetable// presence coeff.)
Toddlers T (0–4, assisted); elderlies E (70+, assisted)	W and H = 1–24; cp = 1 (at home)	W and H = depending on the timetable; cp = 1 for opening times; cp = 0 for closing time
Parents-assisted children PC (5–14, assisted); young autonomous YA (15–19, autonomous)	W = 8–1; cp = 0 (at school)	W and H = depending on the timetable; cp = 1 for opening times; cp = 0 for closing time
	W = 1–7 and 14–24; cp = 1 (at home)	
	H = 1–24; cp = 1 (at home)	
Adults A (20–69, autonomous)	W = 8–18; cp = 0.09 (at work or university)	W and H = depending on the timetable; cp = 1 for opening times; cp = 0 for closing time
	W = 1–7 and 18–24; cp = 1 (at home)	
	H = 1–24; cp = 1 (at home)	

data. Finally, further detail of the age vulnerability distribution can be assessed by considering the percentage of users by health status, by mainly focusing on the frailties outlined in Sect. 2.2.

2.4.4 KPIs for Time-Dependent Assessment

The main KPIs about users' exposure and vulnerability are listed in Table 2.7 [34]. KPIs are derived from NU_t (pp), that is the maximum number of users at a given time t of the day, which is calculated according to Eq. 2.1:

$$NU_t = \sum_{i,a} SU_i \cdot OL_i \cdot cp \cdot AP_a \qquad (2.1)$$

NU_t can be evaluated using hourly sampling, in view of the temporalities described above. Temporalities are considered in view of the combination of the type of areas in the built environment according to Tables 2.4 and 2.5 (subscript i in Eq. 2.1) and the age classes according to Table 2.6 (subscript a). According to the type of analyzed SU_i, the total number of users by their ideal position in the built environment could be also calculated, thus considering the number of only outdoor users $NOOU$, prevalent outdoor users $NPOU$, resident users NRU, and non-resident users $NNRU$ (persons). The same approach could be also applied to specific areas (e.g., to derive the users' number in outdoors $NU_{outdoor,t}$).

In addition to Table 2.7, the overall vulnerability-by-age index VA_t (–) shown in Eq. 2.2 provides a unique KPI to combine the percentage of users by their age (Tp_t,

Table 2.7 Key performance indicators proposed for users' exposure and vulnerability quick characterization, derived from the approach of the BE S²ECURe project [34]

KPI	Symbol (unit of measure)—range	Calculation method	Meaning and utility
Overall users' outdoor density	UOd_t (pp/m^2): 0–3 pp/m^2 (as reasonable maximum condition for overcrowding)	Ratio between NU_t and the overall gross outdoor surface	It expresses the users' density as in the equivalence that all the users are contemporarily placed outdoors. It hence expresses an easy-to-compare value in different scenarios
Users' normalized number	NUn_t (–): 0 (excluded) to 1 (included)	$NU_t/NU_{t,max}$	It compares a given time of the day to the maximum reference conditions in terms of users hosted in the built environment. Thus, $NUn_t = 1$ for the most crowded time of the considered period (e.g., day)
Ration between outdoor and indoor users at a given time t	$OIUr_t$ (–): 0–1	$\sum NU_{outdoor,t}/NU_t$	It evaluates the direct exposure (due to outdoor users) in respect of the overall exposure. Hence, $OIUr_t = 0$ refers to the time of the day in which minimum risk exists, since all the users are indoors, while $OIUr_t = 1$ refers to maximum risk
Percentage of users according to their position	OOp_t, POp_t, Rp_t, NRp_t (%): 0–100%	Ratio between the users by their position and the overall number of users, e.g., $NOOU_t/NU_t$	It investigates vulnerability issues due to the position of the users. Critical values can be retrieved if OOp_t and POp_t are maximized due to their direct exposure to the SLODs

(continued)

2.4 Users' Factors Dynamics

Table 2.7 (continued)

KPI	Symbol (unit of measure)—range	Calculation method	Meaning and utility
Percentage of users for familiarity	$FUUr_t$ (–): 0–1	$\sum(NR_t + OO_t + PO_t)/NU_t$	It investigates vulnerability issues due to familiarity with the built environment, by conservatively considering that only R are familiar with it and its risks. The maximum risk scenario is associated to the max $FUUr_t$
Percentage of users for vulnerability due to age and ideal lack of autonomy in the built environment use	VUr_t (–): 0–1	$\sum(T_t + PC_t + E_t)/NU_t$	Assesses the impact of the most vulnerable users' typologies on the whole number of users in the built environment. The maximum risk scenario is associated to the max VUr_t
Percentage of users for age and gender	$Tp_t, PCp_t, YAp_t, Ap_t, Ep_t, Mp_t, Fp_t$ (%): 0–100%	Directly from census data or as the ration between the number of users by age or gender and the overall number of users	Assesses the impact of vulnerability by each age and gender class. Critical values can be retrieved if values on the most vulnerable users' classes are maximized due to their direct exposure to the SLODs (e.g., T, E, PC)—compare with the principles of VUr_t
Percentage of users by relevant health frailty h	$Hp_{t,h}$ (%): 0–100%	Directly from census data	Assesses the impact of vulnerability by each age and gender class. Critical values can be retrieved if the value is maximized due to their direct exposure to the SLODs. The effects of health frailty on acute and long-lasting SLODs effects can be separately assessed

PCp_t, YAp_t, Ap_t, Ep_t) in respect of their weighted impact, which is derived according to an Analytical Hierarchy Process approach [14]:

$$VA_t = 0.45 \cdot Tp_t + 0.20 \cdot (PCp_t + YAp_t) + 0.04 \cdot Ap_t + 0.31 \cdot Ep_t \qquad (2.2)$$

All these KPIs can rapidly trace differences in users' exposure and vulnerability over the time t, by separately referring to working days and holidays, or event other seasonal or specific built environment use conditions. In this sense, normalization issues could be related to the most critical scenario-based values. Nevertheless, maximum, quartile-based (i.e., at least, median, especially in case of non-Gaussian data distributions) or average values of the KPIs can be assessed, thus regardless of time, so as to have a quick overview of the considered conditions of the public open spaces. If a given KPI in Table 2.7 is calculated referring to statistical values regardless of time, it provides an overview of exposure conditions which can be also compared with other scenarios (i.e., other public open spaces).

References

1. Han S, Song D, Xu L et al (2022) Behaviour in public open spaces: a systematic review of studies with quantitative research methods. Build Environ 223:109444. https://doi.org/10.1016/j.buildenv.2022.109444
2. Zakariya K, Harun NZ, Mansor M (2014) Spatial characteristics of urban square and sociability: a review of the City Square, Melbourne. Procedia Soc Behav Sci 153:678–688. https://doi.org/10.1016/j.sbspro.2014.10.099
3. Paukaeva AA, Setoguchi T, Luchkova VI et al (2021) Impacts of the temporary urban design on the people's behavior—the case study on the winter city Khabarovsk, Russia. Cities 117:103303. https://doi.org/10.1016/j.cities.2021.103303
4. Villagra-Islas P, Alves S (2016) Open space and their attributes, uses and restorative qualities in an earthquake emergency scenario: the case of Concepción, Chile. Urban For Urban Green 19:56–67. https://doi.org/10.1016/j.ufug.2016.06.017
5. Shafray E, Kim S (2017) A study of walkable spaces with natural elements for urban regeneration: a focus on cases in Seoul, South Korea. Sustainability 9:587. https://doi.org/10.3390/su9040587
6. Ge Y, Zhang H, Dou W et al (2017) Mapping social vulnerability to air pollution: a case study of the Yangtze River Delta Region, China. Sustainability 9:109. https://doi.org/10.3390/su9010109
7. Zhang Y, Mao G, Chen C et al (2020) Population exposure to concurrent daytime and nighttime heatwaves in Huai River Basin, China. Sustain Cities Soc 61:102309. https://doi.org/10.1016/j.scs.2020.102309
8. Sabrin S, Karimi M, Fahad MGR, Nazari R (2020) Quantifying environmental and social vulnerability: role of urban heat island and air quality, a case study of Camden, NJ. Urban Clim 34:100699. https://doi.org/10.1016/j.uclim.2020.100699
9. Yang S, Ding L, Prasad D (2022) A multi-sector causal network of urban heat vulnerability coupling with mitigation. Build Environ 226:109746. https://doi.org/10.1016/j.buildenv.2022.109746
10. Villagrán De León JC (2006) Vulnerability: a conceptual and methodological review
11. UNDRR (2016) A/RES/71/644 report of the open-ended intergovernmental expert working group on indicators and terminology relating to disaster risk reduction

References

12. Lin J, Zhu R, Li N, Becerik-Gerber B (2020) How occupants respond to building emergencies: a systematic review of behavioral characteristics and behavioral theories. Saf Sci 122:104540. https://doi.org/10.1016/j.ssci.2019.104540
13. Cheung PK, Jim CY (2019) Improved assessment of outdoor thermal comfort: 1-hour acceptable temperature range. Build Environ 151:303–317. https://doi.org/10.1016/j.buildenv.2019.01.057
14. Salvalai G, Blanco Cadena JD, Sparvoli G et al (2022) Pedestrian single and multi-risk assessment to SLODs in urban built environment: a mesoscale approach. Sustainability 14:11233. https://doi.org/10.3390/su141811233
15. Choi Y, Yoon H, Kim D (2019) Where do people spend their leisure time on dusty days? Application of spatiotemporal behavioral responses to particulate matter pollution. Ann Reg Sci 63:317–339. https://doi.org/10.1007/s00168-019-00926-x
16. Li J, Li J, Yuan Y, Li G (2019) Spatiotemporal distribution characteristics and mechanism analysis of urban population density: a case of Xi'an, Shaanxi, China. Cities 86:62–70. https://doi.org/10.1016/j.cities.2018.12.008
17. Howe PD, Marlon JR, Wang X, Leiserowitz A (2019) Public perceptions of the health risks of extreme heat across US states, counties, and neighborhoods. Proc Natl Acad Sci USA 116:6743–6748. https://doi.org/10.1073/pnas.1813145116
18. O'Lenick CR, Wilhelmi OV, Michael R et al (2019) Urban heat and air pollution: a framework for integrating population vulnerability and indoor exposure in health risk analyses. Sci Total Environ 660:715–723. https://doi.org/10.1016/j.scitotenv.2019.01.002
19. Soomar SM, Soomar SM (2023) Identifying factors to develop and validate a heat vulnerability tool for Pakistan—a review. Clin Epidemiol Glob Health 19:101214. https://doi.org/10.1016/j.cegh.2023.101214
20. Hankey S, Marshall JD, Brauer M (2012) Health impacts of the built environment: within-urban variability in physical inactivity, air pollution, and ischemic heart disease mortality. Environ Health Perspect 120:247–253. https://doi.org/10.1289/ehp.1103806
21. Xiang Z, Qin H, He B-J et al (2022) Heat vulnerability caused by physical and social conditions in a mountainous megacity of Chongqing, China. Sustain Cities Soc 80:103792. https://doi.org/10.1016/j.scs.2022.103792
22. Sabrin S, Karimi M, Nazari R (2022) Modeling heat island exposure and vulnerability utilizing earth observations and social drivers: a case study for Alabama, USA. Build Environ 226:109686. https://doi.org/10.1016/j.buildenv.2022.109686
23. Arsad FS, Hod R, Ahmad N et al (2022) The impact of heatwaves on mortality and morbidity and the associated vulnerability factors: a systematic review. Int J Environ Res Public Health 19:16356. https://doi.org/10.3390/ijerph192316356
24. Birkmann J, Cardona OD, Carreño ML et al (2013) Framing vulnerability, risk and societal responses: the MOVE framework. Nat Hazards 67:193–211. https://doi.org/10.1007/s11069-013-0558-5
25. Deguen S, Amuzu M, Simoncic V, Kihal-Talantikite W (2022) Exposome and social vulnerability: an overview of the literature review. Int J Environ Res Public Health 19:3534. https://doi.org/10.3390/ijerph19063534
26. Kollanus V, Tiittanen P, Lanki T (2021) Mortality risk related to heatwaves in Finland—factors affecting vulnerability. Environ Res 201:111503. https://doi.org/10.1016/j.envres.2021.111503
27. Bălă G-P, Râjnoveanu R-M, Tudorache E et al (2021) Air pollution exposure—the (in)visible risk factor for respiratory diseases. Environ Sci Pollut Res 28:19615–19628. https://doi.org/10.1007/s11356-021-13208-x
28. Wang H, Gao Z, Ren J et al (2019) An urban-rural and sex differences in cancer incidence and mortality and the relationship with PM2.5 exposure: an ecological study in the southeastern side of Hu line. Chemosphere 216:766–773. https://doi.org/10.1016/j.chemosphere.2018.10.183
29. Kecklund L, Andrée K, Bengtson S et al (2012) How do people with disabilities consider fire safety and evacuation possibilities in historical buildings?—A Swedish case study. Fire Technol 48:27–41. https://doi.org/10.1007/s10694-010-0199-0

30. Bosina E, Weidmann U (2017) Estimating pedestrian speed using aggregated literature data. Phys A Stat Mech Appl 468:1–29. https://doi.org/10.1016/j.physa.2016.09.044
31. Kwon K, Akar G (2022) People with disabilities and use of public transit: the role of neighborhood walkability. J Transp Geogr 100:103319. https://doi.org/10.1016/j.jtrangeo.2022.103319
32. Prescott M, Labbé D, Miller WC et al (2020) Factors that affect the ability of people with disabilities to walk or wheel to destinations in their community: a scoping review. Transp Rev 40:646–669. https://doi.org/10.1080/01441647.2020.1748139
33. Hwang J (2022) A factor analysis for identifying people with disabilities' mobility issues in built environments. Transp Res Part F Traffic Psychol Behav 88:122–131. https://doi.org/10.1016/j.trf.2022.05.016
34. Quagliarini E, Bernardini G, Romano G, D'Orazio M (2023) Users' vulnerability and exposure in public open spaces (squares): a novel way for accounting them in multi-risk scenarios. Cities 133:104160. https://doi.org/10.1016/j.cities.2022.104160
35. Dzyuban Y, Ching GNY, Yik SK et al (2022) Outdoor thermal comfort research in transient conditions: a narrative literature review. Landsc Urban Plan 226:104496. https://doi.org/10.1016/j.landurbplan.2022.104496
36. Sharifi A (2019) Urban form resilience: a meso-scale analysis. Cities 93:238–252. https://doi.org/10.1016/j.cities.2019.05.010
37. Lu Y (2023) Drive less but exposed more? Exploring social injustice in vehicular air pollution exposure. Soc Sci Res 111:102867. https://doi.org/10.1016/j.ssresearch.2023.102867
38. Jiang Y, Chen L, Grekousis G et al (2021) Spatial disparity of individual and collective walking behaviors: a new theoretical framework. Transp Res Part D Transp Environ 101:103096. https://doi.org/10.1016/j.trd.2021.103096
39. Shi X, Zheng Y, Cui H et al (2022) Exposure to outdoor and indoor air pollution and risk of overweight and obesity across different life periods: a review. Ecotoxicol Environ Saf 242:113893. https://doi.org/10.1016/j.ecoenv.2022.113893
40. Forehead H, Huynh N (2018) Review of modelling air pollution from traffic at street-level—the state of the science. Environ Pollut 241:775–786. https://doi.org/10.1016/j.envpol.2018.06.019
41. de Nazelle A, Rodríguez DA, Crawford-Brown D (2009) The built environment and health: impacts of pedestrian-friendly designs on air pollution exposure. Sci Total Environ 407:2525–2535. https://doi.org/10.1016/j.scitotenv.2009.01.006
42. Elzeni MM, ELMokadem AA, Badawy NM (2022) Impact of urban morphology on pedestrians: a review of urban approaches. Cities 129:103840. https://doi.org/10.1016/j.cities.2022.103840
43. Akopov AS, Beklaryan LA, Saghatelyan AK (2019) Agent-based modelling of interactions between air pollutants and greenery using a case study of Yerevan, Armenia. Environ Model Softw 116:7–25. https://doi.org/10.1016/j.envsoft.2019.02.003
44. Yang L, Zhang L, Stettler MEJ et al (2020) Supporting an integrated transportation infrastructure and public space design: a coupled simulation method for evaluating traffic pollution and microclimate. Sustain Cities Soc 52:101796. https://doi.org/10.1016/j.scs.2019.101796
45. Falasca S, Ciancio V, Salata F et al (2019) High albedo materials to counteract heat waves in cities: an assessment of meteorology, buildings energy needs and pedestrian thermal comfort. Build Environ 163:106242. https://doi.org/10.1016/j.buildenv.2019.106242
46. Borrego C, Valente J, Amorim JH et al (2012) Modelling of tree-induced effects on pedestrian exposure to road traffic pollution. WIT Trans Built Environ 128:3–13. https://doi.org/10.2495/UT120011
47. Freire S (2010) Modeling of spatiotemporal distribution of urban population at high resolution—value for risk assessment and emergency management. In: Konecny M, Zlatanova S, Bandrova TL (eds) Geographic information and cartography for risk and crisis, pp 53–67
48. Melnikov VR, Krzhizhanovskaya VV, Lees MH, Sloot PMA (2020) The impact of pace of life on pedestrian heat stress: a computational modelling approach. Environ Res 186:109397. https://doi.org/10.1016/j.envres.2020.109397
49. Zhuang L, Huang J, Li F, Zhong K (2022) Psychological adaptation to thermal environments and its effects on thermal sensation. Physiol Behav 247:113724. https://doi.org/10.1016/j.physbeh.2022.113724

References

50. Gonsalves MS, O'Brien B, Twomey DM (2021) Sport and leisure activities in the heat: what safety resources exist? J Sci Med Sport 24:781–786. https://doi.org/10.1016/j.jsams.2021.05.016
51. Ioannou LG, Gkikas G, Mantzios K et al (2021) Risk assessment for heat stress during work and leisure. In: Toxicological risk assessment and multi-system health impacts from exposure. Elsevier, pp 373–385
52. Zabetian E, Kheyroddin R (2019) Comparative evaluation of relationship between psychological adaptations in order to reach thermal comfort and sense of place in urban spaces. Urban Clim 29:100483. https://doi.org/10.1016/j.uclim.2019.100483
53. Wang Y, Eriksson T, Luo N (2023) The health impacts of two policies regulating SO_2 air pollution: evidence from China. China Econ Rev 78:101937. https://doi.org/10.1016/j.chieco.2023.101937
54. Liu J, Jiao J, Xie Y et al (2022) Assessment on the expectation for outdoor usage and its influencing factors. Urban Clim 42:101132. https://doi.org/10.1016/j.uclim.2022.101132
55. Cabanac M (1971) Physiological role of pleasure. Science (80–) 173:1103–1107. https://doi.org/10.1126/science.173.4002.1103
56. Schweiker M, Schakib-Ekbatan K, Fuchs X, Becker S (2020) A seasonal approach to alliesthesia. Is there a conflict with thermal adaptation? Energy Build 212:109745. https://doi.org/10.1016/j.enbuild.2019.109745
57. Yıldız B, Çağdaş G (2020) Fuzzy logic in agent-based modeling of user movement in urban space: definition and application to a case study of a square. Build Environ 169:106597. https://doi.org/10.1016/j.buildenv.2019.106597
58. Semenov A, Zelentsov V, Pimanov I (2019) Application suggesting attractive walking routes for pedestrians using an example of Saint-Petersburg City. Procedia Comput Sci 156:319–326. https://doi.org/10.1016/j.procs.2019.08.208
59. Salazar Miranda A, Fan Z, Duarte F, Ratti C (2021) Desirable streets: using deviations in pedestrian trajectories to measure the value of the built environment. Comput Environ Urban Syst 86:101563. https://doi.org/10.1016/j.compenvurbsys.2020.101563
60. Tong Y, Bode NWF (2022) The principles of pedestrian route choice. J R Soc Interface 19. https://doi.org/10.1098/rsif.2022.0061
61. Filomena G, Kirsch L, Schwering A, Verstegen JA (2022) Empirical characterisation of agents' spatial behaviour in pedestrian movement simulation. J Environ Psychol 82:101807. https://doi.org/10.1016/j.jenvp.2022.101807
62. Kang C-D (2018) The S + 5Ds: spatial access to pedestrian environments and walking in Seoul, Korea. Cities 77:130–141. https://doi.org/10.1016/j.cities.2018.01.019
63. Nolasco-Cirugeda A, García-Mayor C, Lupu C, Bernabeu-Bautista A (2022) Scoping out urban areas of tourist interest though geolocated social media data: Bucharest as a case study. Inf Technol Tour 24:361–387. https://doi.org/10.1007/s40558-022-00235-8
64. Alhazzani M, Alhasoun F, Alawwad Z, González MC (2021) Urban attractors: discovering patterns in regions of attraction in cities. PLoS ONE 16:e0250204. https://doi.org/10.1371/journal.pone.0250204
65. Guo Z, Loo BPY (2013) Pedestrian environment and route choice: evidence from New York City and Hong Kong. J Transp Geogr 28:124–136. https://doi.org/10.1016/j.jtrangeo.2012.11.013
66. Cherfaoui D, Djelal N (2018) Change in use and development of public squares considering daily temporalities. Artic Rev Sci Hum. https://doi.org/10.4000/articulo.3809
67. Quagliarini E, Lucesoli M, Bernardini G (2021) How to create seismic risk scenarios in historic built environment using rapid data collection and managing. J Cult Herit 48:93–105. https://doi.org/10.1016/j.culher.2020.12.007
68. Banerjee A, Maurya AK, Lämmel G (2018) Pedestrian flow characteristics and level of service on dissimilar facilities: a critical review. Coll Dyn 3:A17. https://doi.org/10.17815/CD.2018.17
69. Melnikov V, Krzhizhanovskaya VV, Sloot PMA (2017) Models of pedestrian adaptive behaviour in hot outdoor public spaces. Procedia Comput Sci 108:185–194. https://doi.org/10.1016/j.procs.2017.05.006

70. Briggs DJ, de Hoogh K, Morris C, Gulliver J (2008) Effects of travel mode on exposures to particulate air pollution. Environ Int 34:12–22. https://doi.org/10.1016/j.envint.2007.06.011
71. Chung J, Kim S-N, Kim H (2019) The impact of PM10 levels on pedestrian volume: findings from streets in Seoul, South Korea. Int J Environ Res Public Health 16:4833. https://doi.org/10.3390/ijerph16234833
72. Liang S, Leng H, Yuan Q et al (2020) How does weather and climate affect pedestrian walking speed during cool and cold seasons in severely cold areas? Build Environ 175:106811. https://doi.org/10.1016/j.buildenv.2020.106811
73. Langenheim N, White M, Tapper N et al (2020) Right tree, right place, right time: a visual-functional design approach to select and place trees for optimal shade benefit to commuting pedestrians. Sustain Cities Soc 52:101816. https://doi.org/10.1016/j.scs.2019.101816
74. Xue P, Jia X, Lai D et al (2021) Investigation of outdoor pedestrian shading preference under several thermal environment using remote sensing images. Build Environ 200:107934. https://doi.org/10.1016/j.buildenv.2021.107934
75. de Montigny L, Ling R, Zacharias J (2012) The effects of weather on walking rates in nine cities. Environ Behav 44:821–840. https://doi.org/10.1177/0013916511409033
76. Zare S, Hasheminejad N, Shirvan HE et al (2018) Comparing Universal Thermal Climate Index (UTCI) with selected thermal indices/environmental parameters during 12 months of the year. Weather Clim Extrem 19:49–57. https://doi.org/10.1016/j.wace.2018.01.004
77. Qi J, Wang J, Zhai W et al (2022) Are there differences in thermal comfort perception of children in comparison to their caregivers' judgments? A study on the playgrounds of parks in China's hot summer and cold winter region. Sustainability 14:10926. https://doi.org/10.3390/su141710926
78. Tian Y, Hong B, Zhang Z et al (2022) Factors influencing resident and tourist outdoor thermal comfort: a comparative study in China's cold region. Sci Total Environ 808:152079. https://doi.org/10.1016/j.scitotenv.2021.152079
79. Neset T-S, Navarra C, Graça M et al (2022) Navigating urban heat—assessing the potential of a pedestrian routing tool. Urban Clim 46:101333. https://doi.org/10.1016/j.uclim.2022.101333
80. Siriaraya P, Wang Y, Zhang Y et al (2020) Beyond the shortest route: a survey on quality-aware route navigation for pedestrians. IEEE Access 8:135569–135590. https://doi.org/10.1109/ACCESS.2020.3011924
81. Nurminen A, Malhi A, Johansson L, Framling K (2020) A clean air journey planner for pedestrians using high resolution near real time air quality data. In: Proceedings of 2020 16th international conference on intelligent environments—IE 2020, pp 44–51. https://doi.org/10.1109/IE49459.2020.9155068
82. Sihombing R, Sini SK, Fitzky M (2020) Developing web-based 3D health-aware routing for pedestrians and cyclists. E3S Web Conf 171:02009. https://doi.org/10.1051/e3sconf/202017102009
83. Luo J, Boriboonsomsin K, Barth M (2018) Reducing pedestrians' inhalation of traffic-related air pollution through route choices: case study in California suburb. J Transp Health 10:111–123. https://doi.org/10.1016/j.jth.2018.06.008
84. Choudhary R, Ratra S, Agarwal A (2022) Multimodal routing framework for urban environments considering real-time air quality and congestion. Atmos Pollut Res 13:101525. https://doi.org/10.1016/j.apr.2022.101525
85. García-Palomares JC, Salas-Olmedo MH, Moya-Gómez B et al (2018) City dynamics through Twitter: relationships between land use and spatiotemporal demographics. Cities 72. https://doi.org/10.1016/j.cities.2017.09.007
86. Nemeškal J, Ouředníček M, Pospíšilová L (2020) Temporality of urban space: daily rhythms of a typical week day in the Prague metropolitan area. J Maps 16:30–39. https://doi.org/10.1080/17445647.2019.1709577
87. Cherfaoui D, Djelal N (2019) Assessing the flexibility of public squares the case of Grande Poste square in Algiers. Cities 93:164–176. https://doi.org/10.1016/j.cities.2019.04.017
88. Hahm Y, Yoon H, Choi Y (2019) The effect of built environments on the walking and shopping behaviors of pedestrians; a study with GPS experiment in Sinchon retail district in Seoul, South Korea. Cities 89. https://doi.org/10.1016/j.cities.2019.01.020

References

89. Bloomberg M, Burden A (2006) New York City pedestrian level of service study—phase 1. New York, NY
90. Ministry of Interior (Italy) (2015) DM 03/08/2015: fire safety criteria (Approvazione di norme tecniche di prevenzione incendi, ai sensi dell'articolo 15 del decreto legislativo 8 marzo 2006, n. 139)
91. De Lotto R, Pietra C, Venco EM (2019) Risk analysis: a focus on urban exposure estimation. In: Computational science and its applications—ICCSA 2019. Springer, Cham, pp 407–423

Chapter 3
Quantifying SLODs Risk and Mitigation Potential in Urban BE: A Behavioural Based Approach

Abstract User behavioural response to environmental conditions in the urban Built Environment (BE) can alter the users individual risks to Slow Onset Disasters (SLODs), especially in outdoor scenarios. Herein, micro-climate-related stress, users' vulnerabilities and exposure contributes to the users' reaction and to the susceptibility to SLODs effects on health. A "behavioural-based" approach is proposed and tested in this chapter, merging all of these domains into a simulation perspective and evaluating quantitatively SLOD risk on health. The proposed methodology relies on the virtual simulation of the BE scenario to determine micro-climate conditions, and enable users' behaviour prediction, so as to derive time-dependant users' risk in public open spaces due to heat stress and air pollution stress. Key performance indicators are further proposed by considering the users' stress levels depending on their SLOD-affected distribution in the open spaces and on their individual vulnerabilities. In particular, the developed risk indicators quantify short-term effects such as sweat rate, water loss and health affection rate probability. The method can be used to evaluate current scenarios conditions as well as the implementation of mitigation strategies, which are summarized in this chapter also in relation to their impacts on users' behaviours in outdoors. The overall proposed risk assessment methodology can be then combined to the procedures for analysing users' vulnerability and exposure exposed in Chap. 2, to match together risk factors according to the assumed "behavioural-based" approach and support decision makers in mitigation strategies prioritization and evaluation.

Keywords User behaviours · Heat stress · Air quality · Built environment · Public open spaces · Risk assessment · Key performance indicators · Health risk

3.1 Behavioural-Based Approach Assessment of SLOD Risks in Urban BE

Users' factors can significantly alter the final risk levels due to Slow Onset Disasters (SLODs) in urban Built Environment (BE) and public open spaces, as shown in Chap. 2. Determining "in which manner" and "how long" the BE users are exposed to SLOD-related stress, and more importantly, "how much" their health is influenced by it, means considering how the BE users react to the environmental stressors, also depending on their recurring behaviours in public open spaces fruition [1–3]. Therefore, a "behavioural-based" (or "behavioural design"-based) approach is needed to include users' behaviours as the starting point of the assessment process, to provide user-centered metrics for risk analysis and to support the development of user-oriented mitigation strategies [4]. A "behavioural-based" approach has been widely codified and applied in different BE safety contexts, especially for sudden onset disasters. For instance, Fire Safety Engineering methodologies and the "psychonomics" standpoint at the building scale [5], and of earthquake safety at the urban BE scale [4]. Recent works related to SLODs have adopted the same approach by analysing and simulating probable users' trajectories and distribution outdoor. These have been done in either squares, streets canyons or up to urban districts. And, they have included the analysis of individual stress perception and comfort level depending on the users' choices and actions in the outdoor BE [6–10]. To support the process, simulation methods have been developed starting for experimental data, according to the "behavioural-based" approach criteria.

Nevertheless, only recently researchers have attempted to overcome current limitations of the approach with respect to the single public open spaces scale, that are relevant in view of its attraction rules for users (compare Chap. 2) [11]. A novel quick-to-apply approach for combined heat and air pollution risk assessment in a single public open spaces of urban BE has been developed according to the "behavioural-based" standpoint, by: (1) clearly defining a unified simulation workflow which combines the hazard simulation and the users' behaviours simulation; (2) and defining key performance indicators (KPIs) for SLOD risk assessment based on the effects of these SLODs on the users. The method mainly focuses on the outdoor areas in view of the behavioural interactions (compare Chap. 2) and direct effects of SLODs on the users' health [12–16]. According to the general "behavioural-based" standpoint [4, 8], such an approach can be used to evaluate both the current SLOD risk levels and post-retrofit scenarios and thus to evaluate the best alternatives for the mitigation of single and combined SLOD effects.

Starting from these concepts, Fig. 3.1 summarizes the overall framework pointing out the main operational blocks, derived from consolidated works on the proposed "behavioural-based" standpoint [4]. *Scenario creation* allows to collect data on the "pre-event" scenario in terms of BE features, areas use, climate, users' typologies and behaviours, as well as of possible presence of implemented mitigation strategies. Once the scenario has been modelled, *SLOD simulation* is performed to derived the resulting outdoor micro-climate data that determines spatial heat perception (i.e.,

3.1 Behavioural-Based Approach Assessment of SLOD Risks in Urban BE

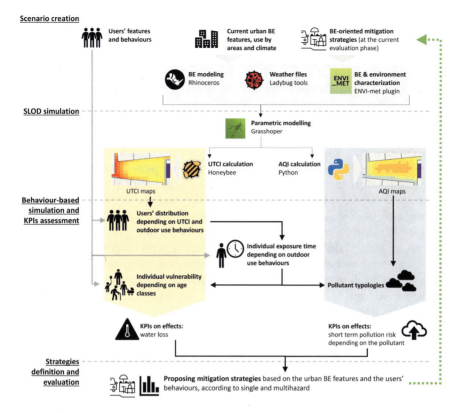

Fig. 3.1 "Behavioural-based" approach for SLOD risk assessment and mitigation in urban BE: operational blocks (on the right) and specific tasks

universal thermal comfort index (UTCI) distribution) and air pollutants concentration (i.e., air quality index (AQI) distribution) [17, 18]. The tasks related to these operational blocks are shown in Sect. 3.2 of this chapter.

Then, *behaviour-based simulations* are performed to populate input data for *KPIs assessment*, as described in following Sect. 3.3. The related tasks focus indeed on the open spaces, that are the outdoor areas, since they are mainly related to critical effects on users during their urban BE fruition [16, 19, 20]. As discussed in Chap. 2, the adopted simulation logic considers that the users' distribution in outdoor areas is strongly linked to the presence of attraction areas and to the spatial distribution of outdoor temperature values (i.e., expressed in local UTCI) [1, 17]. Different exposure times can be evaluated depending on the individual use of public open spaces, by determining the extent of stress from increasing temperatures, thus heat [21], and against air pollutants (according to the type of pollutant) [22, 23]. The resultant health effects are quantified through numerical key performance indicators (KPIs) assessing the water loss level due to the perceived thermal stress, and the short-term air pollution risk. In particular, in view of the individual vulnerability: (1) on heat-related-effects and water loss, it is diversified by age and related body weights; (2) likewise, on air-pollution-related effects and air pollutant typologies, assorted by health affections. Hence, yielding a quick-but-reliable age-class- and health-affection based perspective [11, 24].

In both steps of the framework referring to *SLOD* and *Behaviour-based simulation* (Fig. 3.1), the proposed method adopts quickness criteria aimed at reducing the complexity of scenario definition tasks and, thus, the computation time. In fact, the workflow exploits a granular approach over time and outdoor area spaces [7–9], which is successfully adopted to study users' flows and behaviours in the urban BE at the mesoscale (i.e., for single open spaces and their aggregation) [25, 26]. This approach is combined with simplified-but-robust experimentally-based simulation assumptions concerning the behaviours of users in open spaces fruition, the individual features affecting the users' vulnerability (i.e., by class-based analysis) and the SLOD stress effects on the exposed users' health, as discussed in Chap. 2.

The last step of the workflow refers to the *strategies definition and evaluation* according to the analysis of KPIs and main criticalities emerging from simulations, as discussed in following Sect. 3.4. These strategies could be mainly oriented towards the urban BE, modifying the open spaces features depending on the users' needs and by introducing resilience-increasing components (e.g., furniture, greeneries, surfaces) which can provide benefits from a single and multi-hazard point of view [16, 27–29]. The test of the solutions can be performed again using the overall workflow, and thus it finally compares the KPIs in different pre- and post-intervention scenarios by also taking advantage of successive approximation/steps of analysis (as in the assessment and improvement logic of ISO 14001:2015 [30]).

3.2 From Scenario Creation to SLODs Simulation in Public Open Spaces or Urban BE

The first operational phase of the "behavioural-based" approach in Fig. 3.1 concerns *scenario creation*. This step starts by collecting sufficient data concerning the constitution and morphology of the public open spaces within urban BE, the climate of the specific location, if applicable, eventual previously implemented or planned program of mitigation solutions, and its potential users. The description of the BE materiality and morphology and its relationship with SLOD risk is shown in Chap. 1, Sect. 1.5.1. The assessment of users' factor dynamics and of the related KPIs are presented in Chap. 2, Sect. 2.4. Thus, in this section the attention is instead on scenario descriptors related to the probable distribution of users and resulting micro-climate.

To obtain the results on environmental conditions from computer-based scenario simulations it was required to use Rhinoceros v6[1] + Ladybug Tools v1.6[2] and

---

[1] https://www.rhino3d.com/ (last access: 09/10/2023).
[2] https://www.ladybug.tools/ (last access: 09/10/2023).

ENVIMET v5.[3] In principle, any other 3-Dimensional (3D) CAD software in combination with environmental and substance dispersion modeling tools can be used. These choices allow to retrieve the specific heat stress and air pollution conditions of the given scenario, by detailing the related characteristics into the recreated digital model of the public open spaces.

In greater detail, the BE geometries are virtually recreated in Rhinoceros, and generic radiative and optical properties are allocated to the BE surfaces through the dedicated components of Ladybug Tools. No information on the glazing portion is considered in this method, and it is assumed negligible for the analyzed cases (i.e. historic scenarios). Nevertheless, it can be critical for prevalently glazed and curved facades in the BE. Meanwhile, standards and literature-based values for reflectance are allocated to surfaces representing roofs, facades, pavement and greenery elements (i.e., grass and tree canopies), unless directly surveyed.

The resulting micro-climatic conditions are derived from the environmental modelling process with Ladybug Tools using standardized weather data (i.e., .EPW and .STAT) of the context in which the BE is located. Inputs from in-situ environmental monitoring data are avoided as they can be considered not representative of the overall weather conditions of the context.

Likewise, for air pollutant sources, only standard traffic intensity and generic distribution is included within the evaluated scenario (other local sources are typically uncertain). Thus, vehicle roads are set to be on the perimeter of the public open spaces, considering a concentric pedestrian area (i.e., typical of historic city centers). Traffic congestion level for the roads is selected as medium/high from the default settings on ENVIMET (i.e., 8000 cars/day), prevalently automobiles (i.e., 88%). Regarding background, or pre-existing air pollutants concentration, a more detailed process was required by considering a specific analysis period.

Then, to determine the period for analysing the scenarios in the desired climate, the use of available standard .STAT weather file becomes handy. Such file is used together with dedicated components from Ladybug tools (i.e., LB IMPORT STAT) to rapidly and parametrically individuate the hottest day and week of the representative yearly data. Moreover, considering that the hours of greater solar radiation intensity and traffic rush hour occur between 11:00 and 16:00, it is possible to study the potentially most representative conditions for critical SLOD risk levels (i.e., combining heat stress and air pollution distress).

However, given the complexity and computing expense of the coupled substance dispersion models and Computer Fluid Dynamics (CFD) tools, the distribution of air pollution concentrations (i.e., air pollution distress) were studied at a different time resolution with respect to heat stress. Therefore, the proposed method suggests to consider, for air pollution distress, only values related to daytime from 11:00 till 16:00, during the individuated hottest day of the year. The heat stress parameter results are considered for the same hour timeframe but along the hottest week. Such day is narrowed down by comparing the maximum daily dry-bulb air temperatures on the hottest week of the year.

---

[3] https://www.envi-met.com/ (last access: 09/10/2023).

Having a narrower timeframe to study air pollution, a representative background concentration for each air pollutant (e.g., NO_2, O_3, PM10 and PM2.5) can be obtained by calculating the median air pollution over the hottest week from 11:00 till 16:00. Data can be sourced from available environmental data repositories (see Chap. 1, Sect. 1.3). These values are then set as background concentrations for the computer simulations on ENVIMET.

After setting up the scenarios, the *SLOD simulation* can start. Simulations are executed considering the previously mentioned analysis periods that allow, in a reasonable computing time, to consider more probable conditions for users' exposure to the presented hazards (see Chap. 1, Sects. 1.2, 1.3 and 1.4) rather than an outlier or unrealistic scenario. To this end, a 5 × 5 m analysis grid at 0.9 m height (half of a hypothetically standing 1.80 m-tall user) was set within the public open space, populated with a sensor in the middle of each grid face (different spatial granularity is suggested according to the size of the public open space in analysis). And heat stress rating and air pollution distress indexes (UTCI and AQI respectively) were computed with the following considerations:

- UTCI is obtained using the dedicated Ladybug Tools component (i.e., HB UTCI Comfort Map), assessing approximately the mean radiant temperature. That is, discretized information on the outdoor air temperature, relative humidity and wind speed distribution within the context is not included (CFD calculation for several scenarios are needed to account for it). Instead these were considered equal to the ones provided on standard allocated weather data (.EPW file). Furthermore, longwave radiant temperatures are obtained with Radiance-based spherical view factors in each sensor position, and building surface temperatures obtained from EnergyPlus. Shortwave radiation contributions on each sensor are considered from Radiance-based enhance 2-phase method [31].
- AQI is computed following the guidelines provided by the Environmental Protection Agency (EPA) [32], accounting for the combined effect of diverse air pollutants' concentration present in the discretized public open space grid spaces. In fact, through ENVIMET it was possible to account for particular and gaseous components, as well as sedimentation and deposition process on leaves and surfaces, simulating their photochemical transformation in the NO–NO_2–O_3 reaction cycle.

3.3 Measuring Users' Factors: Behavioural-Based Simulations and KPIs Definition

Users' factors described in Chap. 2 are considered in the whole behavioural-based framework of Fig. 3.1 according to three different levels of assessment and simulation:

3.3 Measuring Users' Factors: Behavioural-Based Simulations and KPIs …   71

1. the users' distribution in the public open spaces, which essentially depends on the UTCI values, and the individual exposure time assessment depending on the outdoor users' behaviours (Sect. 3.3.1);
2. the KPI assessment concerning increasing temperatures, which takes into account the individual vulnerability depending on the users' age, and relies on the estimation of individual water loss (Sect. 3.3.2);
3. the KPI assessment concerning air pollution, which regards the short term pollution risk and considers the typology of pollutants in the BE (Sect. 3.3.3).

The whole simulation and assessment process shown in the following sections focuses on an individual standpoint rather than on the whole group of exposed users, to better highlight which are the microscopic effects and to scale results for different assumed exposure conditions.

3.3.1 Users' Distribution in the Public Open Spaces

According to Fig. 3.2, the users' distribution in the public open spaces is simulated considering that each user could have a given thermal acceptability probability *PA* (%) in respect of the local UTCI values (compare Chap. 2, Sect. 2.3.3). According to an experimental-based analysis [17], two different users' behaviours are considered to calculate PA values, depending on the exposure time in the open space according to the classification shown in Chap. 2, Sect. 2.4.1.

1 h users' behavior can be assigned to Prevalent Outdoor (PO) users, such as customers of dehors, open-air terraces and markets, as well as mass gathering areas, who can remain for a longer time in outdoors depending on their tasks. The PA calculation of PO is shown by Eq. 3.1.

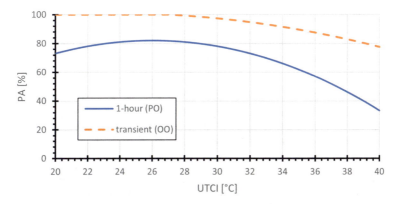

Fig. 3.2 Thermal acceptability probability PA (%) for 1 h (prevalent outdoor—PO) and transient (only outdoor—OO) behaviours, for the UTCI values from 20 to 40 °C

$$PA_{PO} = -0.2485 \cdot UTCI^2 + 12.914 \cdot UTCI - 85.681 \; (\%) \qquad (3.1)$$

Transient users' behavior can be assigned to Only Outdoor (OO) users, such as passersby, who spend less than 15 min in the public open space. The PA calculation of OO is shown by Eq. 3.2.

$$PA_{OO} = -0.0859 \cdot UTCI^2 + 4.019 \cdot UTCI + 54.119 \; (\%) \qquad (3.2)$$

These behaviours relating to the users' presence in outdoors are then based on a granular time representation, which essentially relies on 15 min time-steps. Both users' behaviours can be operatively used to compare the approach sensitivity depending on behavioural factors, in view of their differences in PA values, as shown by Fig. 3.2 in the UTCI range from 20 to 40 °C. In particular, the 1 h thermal acceptability assessment, relating to PO, is considered not to overestimate the acceptability for longer time exposure.

In addition, the acceptability values can be corrected by multiplying the outcoming PA by additional use probability percentages depending on the attractiveness of the single areas in the open spaces for physical, leisure, social and other activities that can be performed depending on the BE temporalities [1, 33] (compare Chap. 2, Sect. 2.3.1). For instance, Eq. 3.3 reduces the PA_{PO} by the percentage of prevalent outdoor users POp_t who can effectively populate the BE at a given time of the day t, as defined by Chap. 2, Sect. 2.4.4. The outcoming probability $PA*_{PO} \leq PA_{PO}$ enables to also consider that other aspects beyond the heat stress on can determine can alter the possibility to remain in a certain part of the outdoor areas relating to PO uses. In general terms, the same approach can be also applied to only outdoor users, to alter passersby's flows depending on the surrounding building use and scheduling.

$$PA^*_{PO} = PA_{PO} \cdot POp_t \; (\%) \qquad (3.3)$$

3.3.2 Heat Stress and Effects on Health

Normally, heat-related impacts are calculated in terms of heat-related mortality and morbidity, expressed in number of reported heat-attributable-deaths increase and expected years of life lost (i.e., YLL) at a large scale (typically city-scale) [34–37]. For instance, Bunker et al. reported, using the PRISMA protocol, that globally a 1 °C temperature rise increased cardiovascular (3.44%, 95% CI 3.10–3.78), respiratory (3.60%, 3.18–4.02), and cerebrovascular (1.40%, 0.06–2.75) mortality [34]. Or, Gasparrini et al. reviewing deaths between 1985 and 2012 in 384 locations around the globe, identifying that 7.71% (95% empirical CI 7.43–7.91) of mortality was attributable to non-optimum temperature [35]. And, Ibbetson et al. who, while analyzing English and Welsh homes during 1 summer time, reported a YLL that varies from 0.49 to 2.58 years depending on the method utilized [36].

3.3 Measuring Users' Factors: Behavioural-Based Simulations and KPIs …

Nevertheless, these analyses are performed at large spatial scale and at an extend time span. Even when scientist have highlighted the importance of rapid and early risk identification, in particular for those socially vulnerable (Chap. 2, Sect. 2.2) [38]. More specific studies have been conducted for children in enclosed spaces (i.e., vehicles) under shadow or direct solar exposure measuring core temperature trends. Extrapolating such results, Vanos et al. reported that a 2 h exposure to this conditions will result in heat injury [39].

In this context, and in order to determine a direct impact on the health of specific public open space users (at a micro-scale which does not necessarily imply residents) at a shorter time span, a different approach was considered. That is, an hourly granular time-dependent effect on people from heat, by working with the potential total sweat driven by the heat stress perceived by the public open space users. Hence, Eq. 3.4 is applied to compute the potential total amount of water loss (WL) in n time steps due to the sweat rate associated to a UTCI-heat stress category, for every time-step t_i (h) defined according to the users' behavioural issues and distribution in Chap. 2, Sect. 2.2. This enables to then compare to any specific group average body weight to see if they are at a significant dehydration risk or not the larger the water loss on body weight rate, the larger the risk (WLR, as in Eq. 3.5).

$$WL = \sum_{i=0}^{n} sweat\,rate_i \cdot t_i \qquad (3.4)$$

$$WLR_{group} = \frac{WL}{body\,weight_{group}} \qquad (3.5)$$

For the UTCI-heat stress allocated sweat rate, information in literature can be utilized or more detailed context-based information can be used as input. For instance, the data provided by Błażejczyk et al. [21, 40] can be used to assign a range of water loss rate for each UTCI heat stress category. Intermediate values were assumed to be linearly interpolated, while sweating rates for UTCI > 46 °C and UTCI < 26 °C are assumed constant. This has been summarized in Table 3.1, as previously presented by Blanco Cadena et al. [11].

Equations 3.4 and 3.5 can be applied in every portion of the discretized area of the placed analysis grid within the public open space. Moreover, as described in Sect. 3.3.1 to account for behavioural based analysis, users can also move in locally critical conditions of UTCI, it can be considered in parallel (or according to the desired typology of user) and continuously for OO and PO. Therefore, for computing an overall WL at the public open space (POS) for a single user or for the whole area, their possible water loss risk in a the analyzed public open space can be determined combining the above mentioned parameters area-weighted averaged following Eq. 3.6 in terms of probability of permanence for each acceptability behavior [17]:

Table 3.1 Sweat rate allocation by UTCI heat stress category, re-elaborated from Błażejczyk et al. [21, 40]

UTCI (°C)	Stress category	Sweat rate (g/h)	Interpolation methods for simulation and estimation
> 46	Extreme heat stress	> 650	Sweat rate = 650
38–46	Very strong heat stress	200–650[a]	$sweat\ rate =$
32–38	Strong heat stress	> 200[a]	$\begin{cases} \frac{225 \cdot UTCI - 5800}{7}, 46 \geq UTCI > 32 \\ \frac{100 \cdot UTCI - 2600}{3}, 32 \geq UTCI > 26 \end{cases}$
26–32	Moderate heat stress	0–200[a]	
9–26	No thermal stress	0	Sweat rate = 0

[a] Undefined. For this work, interpolation values are utilized, using the start and end values of the range as boundary values

$$WL_{POS/group} = \frac{\sum_{a=0}^{n}(Sweat\ rate_a \cdot (t_a \cdot A_a \cdot PA_a))}{\left(\sum_{a=0}^{n} A_a\right) \cdot t_{tot}} \quad (3.6)$$

where a is the area of the discretized grid face, characterized by a given UTCI conditions (and so with a given sweat rate). Each grid face a is characterized by its dimension in surface term is A_a (m^2), and the acceptability PA_a of the UTCI, which varies for OO and PO as reported in Sect. 3.3.1. The reference time is the t_a and t_{tot}, considering the users' permanence in a (as total exposure time) and the POS respectively. Detailed analysis on users' path/presence over the total time within a could be also performed to assess the critical time (t_{crit}) in a, thus in line with Sect. 3.3.1 criteria.

Nevertheless, results can be simplified by focusing on the UTCI effects on users' behaviours. t_a is considered constant and equal to t_{tot}, and individual social vulnerability can be specific (WLR_{group}) depending on age classes. In this case, the groups considered are the ones defined in Chap. 2, Sect. 2.2 but adding a subdivision to the young people category (children and young adults): toddlers (TU), children (PC), young adults (YA), adults (AU), elderly (EU).

The individual physiological conditions by age are considered to be able to specific water loss risk per age group (WLR_{group}), which can also be related to social vulnerability given their individual features (Chap. 2, Sect. 2.2), according to Eq. 3.5 and the information of the mean bodyweight of the population, or group, studied. *Body weight$_{group}$* (g) data can be associated with statistical anthropometric measurements depending on the specific Country in which the analysis is performed. Due to the lack of completeness of available data, information from other relevant databases can be used but acknowledging the potential differences. For instance, the Italian databases are not as granular as the data found for USA demographic statistics[4] (summarized in Table 3.2). As a result, Eq. 3.6 can offer analysis on the water loss, for the assessed POS, in respect to each age group and gender, by additionally considering the impact due to OO and PO behaviours of the related users. In fact, considering only the current application, $WL_{POS/group}$ considers already 20 related values for each assessed POS.

---

[4] https://www.cdc.gov/nchs/data/series/sr_11/sr11_252.pdf (last access: 19/07/2022).

3.3 Measuring Users' Factors: Behavioural-Based Simulations and KPIs ...

Table 3.2 Mean body weight (kg) by representative age classes, and related standard deviation values, according to USA statistics

Age class (years)	Mean body weight—male (std. dev.) (kg)	Mean body weight—female (std. dev.) (kg)
Toddlers TU (< 5)[a]	11 (4)	11 (4)
Parent-assisted children PC (5–14)[a]	40 (14)	40 (14)
Young autonomous users YA (15–18)	77 (4)	65 (2)
Adult users AU (19–65)	89 (3)	76 (1)
Elderly EU (> 65)	83 (4)	70 (5)

[a] Data do not include the weight of the adult moving with the child

3.3.3 Pollution and Effects on Health

Driven by the lack of information on all air pollutant types, the interest on making the methodology robust, and given that no agreement was found on literature on the direct link between short-term exposures of air pollutants and health burden, a slightly different approach was considered. Like what is proposed for heat stress effects on health in Sect. 3.3.2, a more space and time granular analysis is framed to rapidly understand the risk to which public open spaces users are exposed. Thus, the granular pollution burden (i.e., short-term) is estimated as the increase in probability of health affections (reporting symptoms, lung function decrease, asthma, hospital entry and mortality). This approach is similar to that one applied in the work performed by Dockery and Pope for PM10 [23], PM2.5, O_3 and NO_2 by Atkinson et al. [22], and for PM10 and O_3 by Martuzzi et al. [41] where the authors apply the relative risks (RR) for each one of them. Such probability growth or RR are assigned according to the calculated increment of the pollutant concentration (Δpollutant) to which an individual is exposed, compared to air pollutant concentrations equivalent to a target, or ideal, air quality (i.e., AQI).

Thus, as for the heatwave-related risk assessment, a time- and area-weighted parameter is computed (for a single POS user or for the POS itself). In this case, it can be either the concentration of the desired air pollutant per discretized face area, or directly the AQI. Utilizing AQI, Eq. 3.6 is modified into Eq. 3.7, and allows to transform the AQI into the corresponding air pollutant concentration by using Eq. 3.8. From an application perspective, $WL_{POS/group}$ considers 10 related values for each assessed POS, indeed. It must be noted that by using AQI, and not the actual concentration, if AQI was due to a pollutant with a much different behaviour, results might differ from reality. Equation 3.9 hence reports the difference with the suggested and critical concentration values. Finally, the health affection risk (short term pollution risk—STPR) increment can then be determined with Eq. 3.10 utilizing the RR associated to the pollutant analyzed.

$$AQI_{POS/group} = \frac{\sum_{a=0}^{n}(A_a \cdot AQI_a \cdot t_a \cdot PA_a)}{\left(\sum_{a=0}^{n} A_a\right) \cdot t_{tot}} \quad (3.7)$$

$$Concentration_{AQI} = \frac{\left(AQI_{POS/group} - I_{Lo}\right) \cdot (BP_{Hi} - BP_{Lo})}{I_{Hi} - I_{Lo}} + BP_{Lo} \quad (3.8)$$

$$\Delta_{pollutant} = Concentration_{AQI} - Concentration_{suggested} \quad (3.9)$$

$$STPR_i = \left(\frac{\Delta_{pollutant}}{10}\right) \cdot (RR_i - 1) \times 100 \quad (3.10)$$

Referring to the recommendations by Mintz [32], Eqs. 3.7–3.10 consider that: I_{Hi} is the higher bound reference AQI value corresponding to pollutant concentration BP_{Hi}; I_{Lo} is the lower bound reference AQI value corresponding to BP_{Lo}; BP_{Hi} is the breakpoint that is greater than or equal to $Concentration_{AQI}$; and, BP_{Lo} is the breakpoint that is less than or equal to $Concentration_{AQI}$; $Concentration_{suggested}$ can vary according to the goal of the analysis.

In this sense, to report the typology and the extent of the potentially generated health burden in POS users, Eqs. 3.9 and 3.10 can take into account with the information reported by WHO [14] (see Tables 3.3 and 3.4 for different types of pollutants, health burden, and age group classes). The type of health burden hereby presented is focused only on mortality, hospital admissions (cardiovascular), hospital admissions (respiratory) and symptoms relative risk for the specific pollutant type studied.

For estimating the effect of different pollutants in parallel, equivalent emission methods can be applied as well (e.g. for PM10, Foresman et al. [43] has presented a direct method). Nevertheless, these are beyond the scope of this work.

3.4 Mitigation Strategies Definition and Behavioural KPI-Based Evaluation

Mitigation strategies to be tested through the KPI-based assessment, and thus using the "behavioural-based" approach, are resumed in Table 3.5 by including literature references on their definition and application to real-world contexts. Then, their expected separate impacts against effects of increasing temperature and air pollution SLODs, their general applicability and their interactions with users' behaviours are offered in Table 3.6. Both these tables also report codes for the identification of the specific solutions that will be addressed in Chap. 4. All the discussed mitigation strategies are oriented towards a direct implementation in the BE components, to increase their level of implementation in respect of the assessed open space and their users. Mitigation strategies related to water bodies have been excluded, given their localized and low effect, unless unapplicable in dense BE due to large size

3.4 Mitigation Strategies Definition and Behavioural KPI-Based Evaluation

Table 3.3 Type of air pollutants, their associated health affection, and the suggested concentration to mitigate such affections, based on WHO [14], Mintz [32] and US EPA [42]

Pollutant	Health burden type	Suggested concentration	Critical concentration for sensitive groups	Critical concentration for everyone
PM10	Acute lower respiratory infections, cardiovascular disease, chronic obstructive pulmonary disease and lung cancer	15 μg/m³ annual mean 45 μg/m³ 24 h mean	145 μg/m³ 24 h mean[a,b,c]	245 μg/m³ 24 h mean
PM2.5		5 μg/m³ annual mean 15 μg/m³ 24 h mean	40 μg/m³ 24 h mean[a,b,c]	65 μg/m³ 24 h mean
O_3	Asthma morbidity and mortality. Together with reduced lung function and lung diseases	100 μg/m³, 8 h daily maximum (for > 3 days) 60 μg/m³ 8 h mean, peak season (for the most polluted 6 month)	80 μg/m³ 8 h mean[b,c]	105 μg/m³ 8 h mean
NO_2	Asthma, bronchial symptoms, lung inflammation and reduced lung function	10 μg/m³ annual mean 25 μg/m³ 24 h mean 100 μg/m³ 1 h	188 μg/m³ 1 h[b,c]	2354 μg/m³ 1 h[a,b,c]

Sensitive groups can be [a] elderly, [b] children and toddlers, [c] people with health affections (asthma, cardiovascular and respiratory deficiency)

Table 3.4 Collection of the population average short term exposure risk (RR) variance due to an 10 μg/m³ increase on air pollutants concentration. Data collected from Atkinson et al. [22] and Dockery and Pope [23]

Pollutant	Mortality	Hospital admissions (cardiovascular)	Hospital admissions (respiratory)	Symptoms
PM10	1.0100[a]		1.0080[a]	1.028
PM2.5	1.0123	1.0091	1.0190	
O_3	1.0029	1.0089	1.0089	1.0154
NO_2	1.0027	1.0015		

and depth [44–46]. According to previous research outcomes [47],[5] strategies can be distinguished in two main groups, as remarked by Table 3.5 and by the scheme presented in Table 3.3.

---

[5] Also compare the BE S²ECURe project results on "Current BE SLOD risk management and reduction strategies" https://l1nq.com/BES2ECURe-wp2 (last access: 03/08/2023).

Table 3.5 Mitigation strategies and specific solutions oriented to the BE and its components for increasing temperatures and air pollution risk mitigation, including references and real application case studies

Mitigation strategy	CODE: specific solution [References]	Real-world applications [References]
Morphological factors and internal layout of the open spaces		
Improve vegetation	A1: Trees [48–50]	Hamburg [51], Milan [52]
	A2: Shrubs and hedges [53–55]	
	A3: Green barriers [56–58]	
Improve shaded areas	A4: Seasonal shadings [16, 59]	Expo 2015 [60], Metrosol Parasol Seville [61], Umbrella sky Project [62]
Physical and construction factors of open spaces components and of facing buildings		
Outdoor surface temperatures reduction	B1: Urban surface and roughness/cool pavement [63]	Los Angeles white painting project [64]
	B2: Permeable pavers [65]	Derbyshire street [66]
	B3: Permeable grass pavers [65]	Lunix pavers [www.ferrar ibk.it]
Air cleaning solutions	B4: City trees [67]	Green city solutions [67]
Building surface temperatures reduction	B5: Cool façade [68]	Santorini [69]
	B6: Reflective roof/cool roof [70]	White roof project [71]
	B7: Green walls [72–77]	Vegetecture [78], Park lane [79]
	B8: Green roofs [80–88]	Madrid city hall [78], Chicago city hall terrace [89]
Air cleaning solutions	B9: Photocatalytic materials [90–93]	Converse walls [94], Volkswagen walls UK [95]
	B10: Algal pbr [96]	Biq house, Hamburg [96]

Strategies related to "Morphological factors and internal layout of the open spaces" (solutions identified by code A and schematically presented in Fig. 3.3) act on modifications to the geometry and components of the outdoor layout, by introducing elements that do not alter the built fronts configuration but promote the inclusion of greeneries and shading solutions [16, 48, 54]. By this way, Table 3.6 remarks that they can alter the temperature condition and the way in which users can feel discomfort, since they can modify the position of outdoor users' attractors (i.e., for PO) and influence the users' movement in terms of speed and trajectories (both for OO and PO). This result is essentially due to a local decreasing in UTCI levels or to the introduction of potentially shaded areas [6, 17, 97, 98]. Furthermore, some elements having a linear or areal extent, such as barriers and shrubs/hedges, can become relevant obstacles for users' movement, thus additionally altering the general direction for pedestrian flows. In this case, specific analysis to pedestrian

3.4 Mitigation Strategies Definition and Behavioural KPI-Based Evaluation 79

Table 3.6 Estimated effectiveness of mitigation solutions in respect of increasing temperatures and air pollution, including their overall expected potential impact and level of interactions with users according to users' factors and behaviours (compare with Chap. 2). Codes for solutions are reported in Table 3.5

Solution CODE	Effects on increasing temperatures	Effects on air pollution	Applicability	Interaction with users
A1	Air temperature reduction, heat dissipation	Particulate matter dissipation and reduction	+++	Changing attraction areas (i.e. for PO), movement (i.e. speed and trajectories) due to potentially shaded areas
A2	Air temperature reduction[a]	Particulate matter dissipation and reduction	+++	Pedestrian movement (i.e. speed and trajectories) modifications due to the introduction of linear/areal obstacles
A3	Air temperature reduction[a]	Particulate matter dissipation and reduction	+	Pedestrian movement (i.e. speed and trajectories) modifications due to the introduction of linear/areal obstacles
A4	Air temperature reduction, solar radiation reflection	n.a.	++	Changing attraction areas (i.e. for PO), movement and trajectories due to potentially shaded areas
B1	Air temperature reduction, solar radiation reflection, heat dissipation	n.a.	+	Changing attraction areas (i.e. for PO) by potentially altering local UTCI values
B2	Air temperature reduction	n.a.	++	Changing attraction areas (i.e. for PO) by potentially altering local UTCI values
B3	Air temperature reduction, solar radiation reflection	Particulate matter reduction	++	Changing attraction areas (i.e. for PO) by potentially altering local UTCI values
B4	n.a.	Particulate matter dissipation and reduction	+	Possible local obstacles to users' movement
B5	Air temperature reduction, solar radiation reflection, heat dissipation	n.a.	+	Overall possible increased attraction of the open space
B6	Air temperature reduction, solar radiation reflection, heat dissipation	n.a.	++	Overall possible increased attraction of the open space

(continued)

Table 3.6 (continued)

Solution CODE	Effects on increasing temperatures	Effects on air pollution	Applicability	Interaction with users
B7	Air temperature reduction	Particulate matter reduction	+	Overall possible increased attraction of the open space
B8	Air temperature reduction, solar radiation reflection	Particulate matter reduction	++	Overall possible increased attraction of the open space
B9	Solar radiation reflection	Particulate matter reduction	+	n.a.
B10	n.a.	Particulate matter reduction	+	n.a.

[a] Potentially applicable, if integrated with different strategies; n.a.: not assessable or relevant

movement are particularly needed to forecast the probability of significant change of the open spaces performances in normal and emergency conditions [99].

Strategies related to "Physical and construction factors of open spaces components and of facing buildings" (solutions identified by code B) concerns the characteristics of the horizontal and vertical surfaces of building envelopes (roofs and walls) and of open spaces (pavements or vertical elements) [68, 72]. According to Table 3.6, they do not generally alter the users' movement, but can change the position of attraction areas (i.e. for PO) by potentially altering local UTCI values, or even increase

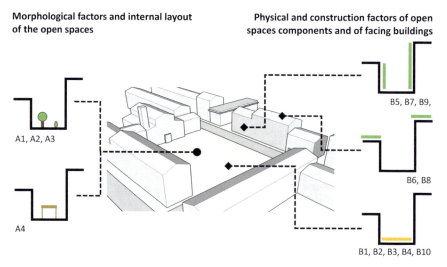

Fig. 3.3 Mitigation strategies applied to the open spaces depending on Table 3.5 classification. Codes for specific solutions are reported in Table 3.6

3.4 Mitigation Strategies Definition and Behavioural KPI-Based Evaluation

the overall attractiveness of the open space in respect to the surrounding urban BE. Table 3.6 also points out the effects on the considered SLODs due to each solution according to the literature works reported in Table 3.5. In this sense, most of the solutions are relevant for one or more issues linked to the single SLOD risk, while some of them (e.g., A1, A2, A3, B3, B7, B8) can have a positive impact on a multi-risk perspective, also in respect of the users' behaviours. Finally, Table 3.6 traces remarks on the solutions applicability, according to a qualitative scale reported by previous research [47]. This scale is roughly divided in three main levels depending on the combination between their implementation according to main literature references, and the frequencies of main retrieved real-world applications: "+" means that the solution is rarely applied/difficulty implementable; "++" means that the solution is often applied/implementable; "+++" means that the solution if very often implemented/easy to implement.

According to the "behavioural-based" approach [4], the KPIs defined in Sect. 3.3.2 can compare different conditions of the given open space to promote the design of mitigation solutions. Thus, they can be employed, for instance, to compare the BE risks before (current state) and after the application of mitigation strategies, once the variation of risk levels for users is quantified. In fact, it is worth noticing that the KPIs are risk indexes, since each of the risk levels grows when each of the KPI grows, and they can separately describe risks for users due to their exposure to increasing temperatures and air pollution effects. The variation of a given KPI can be quantified in absolute terms, as the difference between the a given KPI value in two conditions, and/or in percentage terms, as expressed by $dKPI_{s,r}$ in Eq. 3.11:

$$dKPI_{s,r} = \frac{dKPI_s - dKPI_r}{dKPI_r} \cdot 100 \ (\%) \qquad (3.11)$$

where s refers to the scenario with mitigation strategies implemented in the BE, to be compared with the reference one, expressed by r. As for absolute differences, negative $dKPI_{s,r}$ values imply a decrease of risk, and thus a positive impact of the mitigation strategies.

In addition, the KPIs can be also used to compare different BEs, and thus they can support local administrations to define priorities for interventions, depending on the specific users' risk, in different open spaces in the urban fabric [11]. In this case, differences can be computed in absolute terms, rather than selecting a specific reference open space to apply Eq. 3.11. This will support the priority ranking between the scenarios.

Finally, the proposed KPI-based assessment methodology can be then combined to the procedures for analysing users' vulnerability and exposure exposed in Chap. 2, Sect. 2.4.4, to match together all these risk factors according to the assumed "behavioural-based" approach and support decision makers in mitigation strategies prioritization and evaluation. In this sense, a multirisk metric MR (Eq. 3.12) is proposed to combine conditions of users' exposure and vulnerability EV (%) (Eq. 3.13) to the effects of heat stress HS (–) (Eq. 3.14) and air pollution AP (–) (Eq. 3.15). In Eq. 3.12, both heat stress and air pollution impacts are considered with the same weight, according to a conservative approach in risk assessment [100]. The metric is focused on the most vulnerable users' typology, that are the toddlers, by taking into account their percentage in the scenario TU% (%) (by also assuming this calue as variable over time, thus being equal Tp_t, as shown in Chap. 3, Table 2.7., or assuming the maximum adily value according to a conservative approach when comparing different scenarios). *EV* combines the *TU%* with the density of users for the scenario UOd_t (pp/m²), by normalizing this last value by limit conditions for crowding, that is 3 pp/m² [101] (compare Chap. 2, Table 2.7). *EV* is the associated with: in Eq. 3.13, the heat stress by considering the average WLR$_{group}$ (%) for OO and PO, considering *group* as TU, normalized by the critical WLR value, equal to 4% [40]; in Eq. 3.14, the air pollution impact on the users by the average $STPR_i$ for OO and PO, and both hospitalization and mortality increase. In particular, in Eq. 3.14, *AP* has been normalized by 100% so as to allow the final *MR* being still expressed in percentage terms, varying between 0 and 100%. Maximum values will be associated to highest exposure and presence of TU, given the same *HS* and *AP* conditions.

$$MR = 0.5 \cdot (EV \cdot HS) + 0.5 \cdot (EV \cdot AP) \ (\%) \tag{3.12}$$

$$EV = \text{TU\%} \cdot \frac{UOd_t}{3 \, \text{pp/m}^2} \ (\%) \tag{3.13}$$

$$HS = \frac{(WRL_{TU,OO} + WRL_{TU,PO})/2}{4} \ (-) \tag{3.14}$$

$$AP = \frac{\sum (STPR_{i,OO} + STPR_{i,PO})/2}{100} \ (-) \tag{3.15}$$

References

1. Han S et al (2022) Behaviour in public open spaces: a systematic review of studies with quantitative research methods. Build Environ 223:109444. https://doi.org/10.1016/j.buildenv.2022.109444
2. de Nazelle A et al (2009) The built environment and health: impacts of pedestrian-friendly designs on air pollution exposure. Sci Total Environ 407:2525–2535. https://doi.org/10.1016/j.scitotenv.2009.01.006
3. Andersson-Sköld Y et al (2015) An integrated method for assessing climate-related risks and adaptation alternatives in urban areas. Clim Risk Manag 7:31–50. https://doi.org/10.1016/j.crm.2015.01.003
4. Bernardini G et al (2016) Towards a "behavioural design" approach for seismic risk reduction strategies of buildings and their environment. Saf Sci 86:273–294. https://doi.org/10.1016/j.ssci.2016.03.010
5. Kobes M et al (2010) Building safety and human behaviour in fire: a literature review. Fire Saf J 45:1–11. https://doi.org/10.1016/j.firesaf.2009.08.005

References

6. Yıldız B, Çağdaş G (2020) Fuzzy logic in agent-based modeling of user movement in urban space: definition and application to a case study of a square. Build Environ 169. https://doi.org/10.1016/j.buildenv.2019.106597
7. Abdallah ASH (2022) Urban morphology as an adaptation strategy to improve outdoor thermal comfort in urban residential community of new Assiut City, Egypt. Sustain Cities Soc 78:103648. https://doi.org/10.1016/j.scs.2021.103648
8. Estacio I et al (2022) Optimization of tree positioning to maximize walking in urban outdoor spaces: a modeling and simulation framework. Sustain Cities Soc 86:104105. https://doi.org/10.1016/j.scs.2022.104105
9. Mortezazadeh M et al (2021) Integrating CityFFD and WRF for modeling urban microclimate under heatwaves. Sustain Cities Soc 66:102670. https://doi.org/10.1016/j.scs.2020.102670
10. Engel MS et al (2018) Perceptual studies on air quality and sound through urban walks. Cities 83. https://doi.org/10.1016/j.cities.2018.06.020
11. Blanco Cadena JD et al (2023) Determining behavioural-based risk to SLODs of urban public open spaces: key performance indicators definition and application on established built environment typological scenarios. Sustain Cities Soc 95:104580. https://doi.org/10.1016/j.scs.2023.104580
12. Gherri B et al (2021) On the thermal resilience of Venetian open spaces. Heritage 4:4286–4303. https://doi.org/10.3390/heritage4040236
13. World Health Organization (WHO) (2016) Ambient air pollution: a global assessment of exposure and burden of disease. World Health Organization
14. World Health Organization (2021) Ambient (outdoor) air pollution. https://www.who.int/news-room/fact-sheets/detail/ambient-(outdoor)-air-quality-and-health. Accessed 25 Oct 2021
15. Krüger EL et al (2017) Short- and long-term acclimatization in outdoor spaces: exposure time, seasonal and heatwave adaptation effects. Build Environ 116:17–29. https://doi.org/10.1016/j.buildenv.2017.02.001
16. Paolini R et al (2014) Assessment of thermal stress in a street canyon in pedestrian area with or without canopy shading. Energy Procedia 48:1570–1575. https://doi.org/10.1016/j.egypro.2014.02.177
17. Cheung PK, Jim CY (2019) Improved assessment of outdoor thermal comfort: 1-hour acceptable temperature range. Build Environ 151:303–317. https://doi.org/10.1016/j.buildenv.2019.01.057
18. Fang Z et al (2011) A proposed pedestrian waiting-time model for improving space–time use efficiency in stadium evacuation scenarios. Build Environ 46:1774–1784. https://doi.org/10.1016/j.buildenv.2011.02.005
19. Choi Y et al (2019) Where do people spend their leisure time on dusty days? Application of spatiotemporal behavioral responses to particulate matter pollution. Ann Reg Sci 63:317–339. https://doi.org/10.1007/s00168-019-00926-x
20. Blanco Cadena JD et al (2021) A new approach to assess the built environment risk under the conjunct effect of critical slow onset disasters: a case study in Milan, Italy. Appl Sci 11:1186. https://doi.org/10.3390/app11031186
21. Błażejczyk K et al (2010) Principles of the new Universal Thermal Climate Index (UTCI) and its application to bioclimatic research in European scale. Misc Geogr 14:91–102
22. Atkinson R et al (2013) Health risks of air pollution in Europe—HRAPIE project. Recommendations for concentration–response functions for cost–benefit analysis of particulate matter, ozone and nitrogen dioxide. Copenhagen
23. Dockery DW, Pope CA (1994) Acute respiratory effects of particulate air pollution. Annu Rev Public Health 15:107–132
24. Blanco Cadena JD et al (2021) Flexible workflow for determining critical hazard and exposure scenarios for assessing SLODs risk in urban built environments. Sustainability 13. https://doi.org/10.3390/su13084538
25. Jiang F et al (2022) Pedestrian volume prediction with high spatiotemporal granularity in urban areas by the enhanced learning model. Sustain Cities Soc 79:103653. https://doi.org/10.1016/j.scs.2021.103653

26. Jardim B et al (2023) A street-point method to measure the spatiotemporal relationship between walkability and pedestrian flow. Comput Environ Urban Syst 104:101993. https://doi.org/10.1016/j.compenvurbsys.2023.101993
27. Tran PTM et al (2022) Nature-based solution for mitigation of pedestrians' exposure to airborne particles of traffic origin in a tropical city. Sustain Cities Soc 87:104264. https://doi.org/10.1016/j.scs.2022.104264
28. Sodoudi S et al (2018) The influence of spatial configuration of green areas on microclimate and thermal comfort. Urban For Urban Green 34:85–96. https://doi.org/10.1016/j.ufug.2018.06.002
29. FAO (2016) Building greener cities: nine benefits of urban trees
30. British Standard Institution (BSI) (2015) BS EN ISO 14001: environmental management systems—requirements with guidance for use
31. Ladybug-tools pollination—platform for environmental simulation. https://www.pollination.cloud/
32. Mintz D (2006) Guidelines for the reporting of daily air quality—air quality index (AQI). United States Environmental Protection Agency, Washington
33. Quagliarini E (2022) Users' vulnerability and exposure in public open spaces (squares): a novel way for accounting them in multi-risk scenarios. Cities 133:104160
34. Bunker A et al (2016) Effects of air temperature on climate-sensitive mortality and morbidity outcomes in the elderly; a systematic review and meta-analysis of epidemiological evidence. EBioMedicine 6:258–268. https://doi.org/10.1016/j.ebiom.2016.02.034
35. Gasparrini A et al (2015) Mortality risk attributable to high and low ambient temperature: a multicountry observational study. Lancet 386:369–375. https://doi.org/10.1016/S0140-6736(14)62114-0
36. Ibbetson A et al (2021) Mortality benefit of building adaptations to protect care home residents against heat risks in the context of uncertainty over loss of life expectancy from heat. Clim Risk Manag 32:100307. https://doi.org/10.1016/j.crm.2021.100307
37. CEM-DAT (2023) EM-DAT: the OFDA/CRED international disaster database. Centre for Research on the Epidemiology of Disasters, Universidad Católica de Lovaina, Bruselas
38. Hoffman JL (2001) Heat-related illness in children. Clin Pediatr Emerg Med 2:203–210. https://doi.org/10.1016/S1522-8401(01)90006-0
39. Vanos JK et al (2018) Evaluating the impact of solar radiation on pediatric heat balance within enclosed, hot vehicles. Temperature 5:276–292. https://doi.org/10.1080/23328940.2018.1468205
40. Błażejczyk K et al (2014) Heat stress and occupational health and safety—spatial and temporal differentiation. Misc Geogr Reg Stud Dev 18:61–67. https://doi.org/10.2478/mgrsd-2014-0011
41. Martuzzi M et al (2006) Health impact of PM10 and ozone in 13 Italian cities
42. US EPA (2018) Technical assistance document for the reporting of daily air quality—the air quality index (AQI)
43. Foresman EL et al (2003) PM10 conformity determinations: the equivalent emissions method. Transp Res Part D Transp Environ 8:97–112
44. Liu Z et al (2021) Heat mitigation benefits of urban green and blue infrastructures: a systematic review of modeling techniques, validation and scenario simulation in ENVI-met V4. Build Environ 200:107939. https://doi.org/10.1016/j.buildenv.2021.107939
45. Jacobs C et al (2020) Are urban water bodies really cooling? Urban Clim 32:100607. https://doi.org/10.1016/j.uclim.2020.100607
46. Balany F et al (2022) Studying the effect of blue-green infrastructure on microclimate and human thermal comfort in Melbourne's central business district. Sustainability 14:9057. https://doi.org/10.3390/su14159057
47. Rosso F et al (2023) Mitigating multi-risks in the historical built environment: a multi-strategy adaptive approach. In: Littlewood J, Howlett RJ, Jain LC (eds) Sustainability in energy and buildings 2022. SEB 2022. Smart innovation, systems and technologies, vol 336. Springer, Singapore, pp 197–207

References

48. Janhäll S (2015) Review on urban vegetation and particle air pollution—deposition and dispersion. Atmos Environ
49. Abhijith KV et al (2017) Air pollution abatement performances of green infrastructure in open road and built-up street canyon environments—a review. Atmos Environ 162:71–86. https://doi.org/10.1016/j.atmosenv.2017.05.014
50. Abhijith KV, Gokhale S (2015) Passive control potentials of trees and on-street parked cars in reduction of air pollution exposure in urban street canyons. Environ Pollut. https://doi.org/10.1016/j.envpol.2015.04.013
51. Hamburg.de Grünes Netz
52. Pastore MC (2020) ForestaMi. In: For Rep 2020
53. Gromke C et al (2016) Influence of roadside hedgerows on air quality in urban street canyons. Atmos Environ. https://doi.org/10.1016/j.atmosenv.2016.05.014
54. Gallagher J (2015) Passive methods for improving air quality in the built environment: a review of porous and solid barriers. Atmos Environ
55. Van Ryswyk K et al (2019) Does urban vegetation reduce temperature and air pollution concentrations? Findings from an environmental monitoring study of the Central Experimental Farm in Ottawa, Canada. Atmos Environ. https://doi.org/10.1016/j.atmosenv.2019.116886
56. Al-Dabbous AN, Kumar P (2014) The influence of roadside vegetation barriers on airborne nanoparticles and pedestrians exposure under varying wind conditions. Atmos Environ. https://doi.org/10.1016/j.atmosenv.2014.03.040
57. Yin S (2007) Effects of vegetation status in urban green spaces on particles removal in a canyon street atmosphere. Shengtai Xuebao/Acta Ecol Sin. https://doi.org/10.1016/s1872-2032(08)60007-4
58. Chen X et al (2015) Efficiency differences of roadside greenbelts with three configurations in removing coarse particles (PM10): a street scale investigation in Wuhan, China. Urban For Urban Green. https://doi.org/10.1016/j.ufug.2015.02.013
59. Hwang RL et al (2011) Seasonal effects of urban street shading on long-term outdoor thermal comfort. Build Environ 46:863–870. https://doi.org/10.1016/j.buildenv.2010.10.017
60. Majowiecki M (2015) Roof of the pedestrian walkway for the Milan Expo 2015 exhibition complex. In: Majowiecki website
61. Casella B (2018) Il Metropol Parasol simbolo del rapporto tra tradizione e innovazione. Civiltà di Cantiere, Città e Territ
62. Sampson H (2021) Umbrella sky project: how a Mary Poppins-inspired installation became a tourist magnet. Washington Post
63. Synnefa A (2007) Cool-colored coatings fight the urban heat-island effect. SPIE Newsroom. https://doi.org/10.1117/2.1200706.0777
64. Abdallah C (2018) Los Angeles is painting the streets white (again), and your city might be next. ArchDaily
65. Bell R (2008) Reducing urban heat islands: compendium of strategies trees and vegetation
66. Susdrain (2014) Case study, Derbyshire Street Pocket Park, London Borough of Tower Hamlets
67. Green City Solutions (2021) CITYTREE. Green City Solutions
68. EPA (2008) Urban heat island basics. In: Reducing urban heat islands: compendium of strategies. Heat Island Effect l US EPA
69. Akbari H (2006) Cool color roofing materials. Draft final report prepared for energy 73
70. Falasca S, Curci G (2018) Impact of highly reflective materials on meteorology, PM10 and ozone in urban areas: a modeling study with WRF-CHIMERE at high resolution over Milan (Italy). Urban Sci 2:18. https://doi.org/10.3390/urbansci2010018
71. Kptecki P (2018) New York City painted over 9.2 million square feet of rooftop white. In: Insider
72. Vox G et al (2017) Green façades to control wall surface temperature in buildings. Build Environ. https://doi.org/10.1016/j.buildenv.2017.12.002
73. Manso M, Castro-Gomes J (2015) Green wall systems: a review of their characteristics. Renew Sustain Energy Rev 41:863–871. https://doi.org/10.1016/j.rser.2014.07.203

74. Viecco M et al (2018) Potential of particle matter dry deposition on green roofs and living walls vegetation for mitigating urban atmospheric pollution in semiarid climates. Sustainability. https://doi.org/10.3390/su10072431
75. Ottelé M et al (2010) Quantifying the deposition of particulate matter on climber vegetation on living walls. Ecol Eng. https://doi.org/10.1016/j.ecoleng.2009.02.007
76. Kohler M (2006) Living wall systems—a view back and some visions. In: 4th annual international greening rooftops for sustainable communities conference, awards and trade show
77. Köhler M (2008) Green facades—a view back and some visions. Urban Ecosyst. https://doi.org/10.1007/s11252-008-0063-x
78. Caro JC (2016) Madrid + natural: nature-based climate change adaptation
79. Sue-feng T. Living walls: horticultural wonders in the concrete jungle—台灣光華雜誌
80. Shafique M et al (2018) Green roof benefits, opportunities and challenges—a review. Renew Sustain Energy Rev 90:757–773. https://doi.org/10.1016/j.rser.2018.04.006
81. Sun T et al (2016) How do green roofs mitigate urban thermal stress under heat waves? J Geophys Res. https://doi.org/10.1002/2016JD024873
82. Karachaliou P et al (2016) Experimental and numerical analysis of the energy performance of a large scale intensive green roof system installed on an office building in Athens. Energy Build. https://doi.org/10.1016/j.enbuild.2015.04.055
83. Pérez G (2015) The thermal behaviour of extensive green roofs under low plant coverage conditions. Energy Effic. https://doi.org/10.1007/s12053-015-9329-3
84. EPA (2019) Using green roofs to reduce heat islands. In: Heat Islands
85. Imran HM et al (2018) Effectiveness of green and cool roofs in mitigating urban heat island effects during a heatwave event in the city of Melbourne in southeast Australia. J Clean Prod 197:393–405. https://doi.org/10.1016/j.jclepro.2018.06.179
86. Berardi U et al (2014) State-of-the-art analysis of the environmental benefits of green roofs. Appl Energy
87. Yang J et al (2008) Quantifying air pollution removal by green roofs in Chicago. Atmos Environ. https://doi.org/10.1016/j.atmosenv.2008.07.003
88. Speak AF et al (2012) Urban particulate pollution reduction by four species of green roof vegetation in a UK city. Atmos Environ
89. MWH (2004) Green roof test plot 2003 end of ELD project summary report
90. Mo J et al (2009) Photocatalytic purification of volatile organic compounds in indoor air: a literature review. Atmos Environ 43:2229–2246. https://doi.org/10.1016/j.atmosenv.2009.01.034
91. Pelaez M (2012) A review on the visible light active titanium dioxide photocatalysts for environmental applications. Appl Catal B Environ
92. Kolarik J, Toftum J (2012) The impact of a photocatalytic paint on indoor air pollutants: sensory assessments. Build Environ 57:396–402. https://doi.org/10.1016/j.buildenv.2012.06.010
93. Enea D et al (2010) Photocatalytic properties of cement-based plasters and paints containing mineral pigments. Transp Res Rec. https://doi.org/10.3141/2141-10
94. Elassar A (2020) Converse is sponsoring giant murals that break down air pollutants in 13 cities around the world. In: CNN style
95. Casalgrande Padana (2020) Self cleaning. Casalgrande Padana
96. Loomans T (2013) The world's first algae-powered building opens in Hamburg. Inhabitat
97. Liang S et al (2020) How does weather and climate affect pedestrian walking speed during cool and cold seasons in severely cold areas? Build Environ 175:106811. https://doi.org/10.1016/j.buildenv.2020.106811
98. Melnikov VR et al (2020) The impact of pace of life on pedestrian heat stress: a computational modelling approach. Environ Res 186:109397. https://doi.org/10.1016/j.envres.2020.109397
99. Quagliarini E et al (2023) How could increasing temperature scenarios alter the risk of terrorist acts in different historical squares? A simulation-based approach in typological Italian squares. Heritage 6:5151–5188. https://doi.org/10.3390/heritage6070274

References

100. Salvalai G et al (2022) Pedestrian single and multi-risk assessment to SLODs in urban built environment: a mesoscale approach. Sustainability 14:11233. https://doi.org/10.3390/su141811233
101. Bloomberg M (2006) New York City pedestrian level of service study—phase 1. New York, NY

Chapter 4
Applications to Case Studies

Abstract This chapter offers the application of the "behavioural-based" approach for Slow Onset Disasters (SLODs) assessment and mitigation strategies testing, as proposed by the current work. Two different case studies are used to this end, referring to public squares in historical urban scenarios, and by unfolding the health risk posed by heat stress and air pollution distress derived from the generated microclimate within a single open space in the urban Built Environment (BE). The first case study is a representative typological public open space, based on the analysis of real-world squares in the Italian historical cities. Being an archetype of public open spaces, the application provide general insights on SLODs risks and on the capabilities of mitigation strategies, which can refer to a variety of real-world scenarios sharing the same main features. The second case study is a real-world square in a historical city in Italy, to better understand how general insights can be tailored in comparison to the peculiarities of a given scenario. In both cases, the application follows the proposed process concerning data collection (about SLODs features and related hazard conditions, BE vulnerability, users' exposure and individual vulnerability), simulation scenario definition, simulation-based assessment of pre-retrofit (current) risk levels and simulation-based evaluation of mitigation strategies effectiveness. Indicators such as occupancy probability, thermal perception stress levels, and air quality distress are mapped to determine local hot spots within the studied scenarios. Then, key performance indicators are estimated on SLOD-affected-health potential in the open spaces and on their individual vulnerabilities. The method is expected to enhance the municipalities capacity of resilience towards climate change, by accelerating risk assessments, mitigation strategies planning and execution.

Keywords User behaviours · Heat stress · Air quality · Built environment · Public open spaces · Risk assessment · Digital models · Typological built environment · Health risk

4.1 Typological and Real World Scenarios for the Application of the "Behavioural-Based" Approach on SLOD Risk Assessment

The application of the "behavioural-based" approach to case studies is based on the methods and tools defined in the previous chapters, so as to demonstrate the capabilities of the proposed assessment methods discussed in Chaps. 2 and 3, and then to provide preliminary effectiveness assessment of Slow-Onset Disasters (SLODs) mitigation strategies defined in Chap. 3.

To this end, the first step concerns the definition of the scenarios from a Built Environment (BE)-oriented perspective. In particular, the two case studies considered in this chapter represent squares in historical urban scenarios, as significant open spaces in the urban BE. In fact, historical BEs have been selected because of their peculiar features increasing risks for users in respect to the Slow-Onset Disasters (SLODs) of interest for this work (air pollution and increasing temperature leading to heatwaves), correlated to the physical vulnerability, the used materials, the open space morphology and dimension in respect to the surrounding compact urban fabric, as well as to users' exposure and individual vulnerability in view of the attraction of citizens and visitors [1–6].

The first case study concerns a typological public open space which represents an archetype of an historical square, in terms of morphology, materiality, users' exposure and vulnerability. In fact, previous works [6, 7] collected such data from a wide dataset of real-world squares (more than 1000 sample) referred to Italian historical cities prone to SLODs, by then applying a cluster-based approach to identify Built Environment Typology (BET) characterized by similar recurring conditions. Each BET is hence described by specific statistical-based values (i.e., median and quartiles to characterized also non-normal distributions). A similar typological approach has been widely adopted to investigate urban BEs, including the context of air pollution and heat stress [8–11], and it can be used to provide general insights on SLODs risks and on the capabilities of mitigation strategies, which can refer to a variety of real-world scenarios sharing the same main features. Although BETs are actually based on Italian squares, the methodology could be adopted for other kinds of urban open spaces and BE, as well as in other contexts (e.g., geographical, climatic, social, morphological). In particular, within the whole reference BETs defined in the reference work [7], the considered BET (called BET2 in the reference work) is characterized by critical features in terms of potential multi-risk. In fact, concerning morphology, the low level of compactness and regularity, with a high ratio between the outdoor spaces' width and the facing buildings, can limit the shaded areas in view of increasing temperature can determine critical conditions. The presence of an historical building open to the public acts as an attractor for users (see Chap. 2), thus potentially increasing users' exposure (number of exposed people) and vulnerability (different typologies of people) in respect to other squares without the same intended

4.1 Typological and Real World Scenarios for the Application ...

use. In addition, there is no presence of greenery elements which can potentially mitigate both increasing temperatures and air pollution SLODs effects on the outdoor space users.

Beside this idealized scenario, the second case study refers to a real-world square in a historical city in Italy, that is Piazza dei Priori, in Narni, Italy (N42.51, E12.51). This significant urban open space is prone to the considered SLODs in view of its geographical and climate context, and shares similar main typological features with BET2 [7] (i.e., the low level of compactness and regularity, the presence of buildings open to public, and the absence of greenery elements), although specific conditions in the real-world scenario parameters imply differences with the BET2. In this sense, the application to this real-world scenario allows to better understand how general insights from BETs can be tailored depending on the peculiarities of a given real scenario, that refer to morphological, materials-related and users-related features. Proving the utility of such a method to simplify and promote similar studies in different areas without an expeditive surveying campaign to recreate the real scenario. Furthermore, the application still enables demonstrating the capabilities of the methods in view of decision-makers' actions on risk assessment and mitigation. In general terms, as described in Chap. 3, Sect. 3.2, different inputs are required to accurately describe both the typological and real-world case studies before any computer-aided simulation analysis can be performed. These inputs are extracted from the environmental context, the public open spaces or urban BE, and the SLOD related information (see Fig. 1.11 in Chap. 1), regarding morphology (i.e., geometry), climate, surface materials, vegetation types, background pollution and pollutants sources. In addition, a virtual analysis sensor grid needs to be introduced to better map the conditions, this is normally set as a squared-face grid with regular spacing, unless the area of analysis is rather asymmetrical. In such cases, the reference space geometry can be deconstructed into both squared and triangulated faces to cover sharp edges. A summary of the required inputs is provided in Table 4.1.

Geometrical data is needed to generate digital models in 3D CAD based software (e.g., Rhinoceros,[1] compare Fig. 3.1 in Chap. 3). Having a definition of context-based BETs allows to generalize and speed up the process for a particular region or nation, in particular if these are offered already in a 3D CAD format [7]. On the contrary, in real-world scenarios, these data can be retrieved by on-site measurements (e.g., topography, laser scan), by accessing available georeferenced data on open repositories (e.g., municipality's shapefiles, OpenStreetMap,[2] EUBUCCO[3]) and/or by directly measuring and drawing on scale images (e.g., Google Maps[4]) which allows to describe the BE morphology in terms of footprint and elevation. Top views of the virtual models, including buildings heights, for both the typological and the real-world case study is proposed in Fig. 4.1, considering that both of them have a

---

[1] https://www.rhino3d.com/ (last access: 09/10/2023).
[2] https://www.openstreetmap.org (last access: 23/10/2023).
[3] https://eubucco.com/ (last access: 23/10/2023).
[4] https://www.google.com/maps (last access: 23/10/2023).

Table 4.1 Summary of required input data to map the SLODs hazard risk element for the proposed "behavioral-based" approach for typological or real-world case scenarios

Input type	Parameter	Description
Morphological (geometrical)	x–y projected shape	Buildings, streets, and any significant elements of the public open spaces or urban BE that can highly influence the wind flow and radiation bounces within the outdoor space should be modeled in a coarse way
	z location and height	
	Geolocation	
	Orientation	
	Grid analysis shape	Analysis grid reference surface, location, and space resolution should cover the areas of interest. More than one grid is suggested, if separated calculations are needed
	Grid analysis z location	
	Grid size and spacing	
Climatic	Climate class	The classification is based on threshold limits on either temperature, precipitation and/or related indexes (e.g., degree days)
	Dry-bulb air temperature	General weather data can describe and compute derived micro-climate conditions. These are normally representative conditions of a typical year or of the most representative year for a defined timespan. For the analysis, only the data for the defined analysis period should be utilized
	Relative humidity	
	Atmospheric pressure	
	Solar irradiation (direct, diffuse, and total)	
	Wind speed and direction	
	Prevalent wind speed and direction	Median wind speed and mode of the wind direction, during the analysis period, is set as fixed for air dispersion modelling
	Hottest week	The time period of analysis in simulations can be derived from statistical analysis for the given year
	Hottest day	
Materiality	Surface albedo (reflectivity)	It defines the dynamics of solar radiation and heat transfer within the modelled public open spaces or urban BE
	Surface transmissivity	
	Surface transmittance	
	Density	
	Heat capacity	
Vegetation	Leaf area index	It determines the degree of shading and solar radiation shielding. The specific parameter can be taken from literature and default software values
	Height	
Pollution	Background pollution	Starting pollution conditions in every portion of the model are considered to perform reliable air pollutant dispersion analysis. Such conditions are set as the median values for the analysis periods for the previous 5 years from the date of the analysis
	Pollutants sources	Set from the modelling tools, they are required to understand their effects and the concentration of air pollution. It requires conscious assumptions (e.g., for traffic, intensity, vehicle type distribution, number of lanes) based on the context analyzed
	Pollutant emission rate profiles	

4.1 Typological and Real World Scenarios for the Application … 93

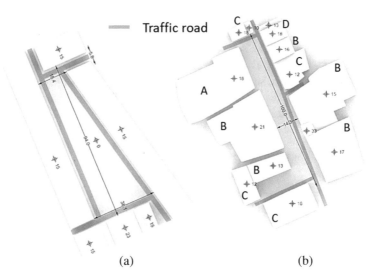

Fig. 4.1 Resulting virtual models top view of **a** BET2 and **b** Piazza dei Priori in Narni. Measurements are all in meters. The north is upwards, and the square ground level is flat. Carriageways have been highlighted by grey lines, as well as building roof-façade combination types by letters

332° azimuth from north (clockwise rotation) for the main dimension of the public open space to make the BET2 consistent with the orientation of Piazza dei Priori.

The analysis grid is set to cover the public open space only (excluding street canyons), to be both quadratic and triangular, with a 5 m maximum spacing between sensor nodes, and at a 0.9 m from the ground (assumed as the approximate centroid of a person's body of 1.75 m height). Detailed analysis can be computed for different heights according to the context or the specific target group of the analysis.

Climatic data is required to be able to estimate micro-climatic conditions in the areas of interest. In this case, the same data are considered for both BET2 and Piazza dei Priori. Environmental data can be found from weather stations on either national, regional or local weather data repositories. However, for most environmental modeling software, these are needed in standardized formats, such as .EPW.[5] These standard files are constructed to have both extreme and most recurrent weather data for a particular site, based on statistical analysis of long period data sets (e.g., 30 years). Such weather files can be found on open dedicated repositories (e.g., EnergyPlus,[6] or Onebuilding[7]) or sourced from proprietary software (e.g., Meteonorm[8]). They include a climate or weather category according to national (e.g., Koppen

---

[5] https://bigladdersoftware.com/epx/docs/8-3/auxiliary-programs/weather-format-for-simulation-programs.html#weather-format-for-simulation-programs (last access: 23/10/2023).

[6] https://energyplus.net/weather (last access: 26/10/2023).

[7] https://climate.onebuilding.org/ (last access: 26/10/2023).

[8] https://meteonorm.com/en/ (last access: 23/10/2023).

Geiger [12]) or international (e.g., ASHRAE[9]) guidelines, depending on the origin or author of the file. And for a generalized analysis, using typological urban BE units, it could be possible to assess the risk with representative cities and climate categories of the context of interest. For the specific case study and for the BET analysis, the weather data on Narni in the form of an .EPW file is used (obtained with Meteonorm). For which, following the Koppen Geiger scheme, a Temperate—hot summer—no dry season climate category is given (Cfa).

Profiting from the .STAT file for the weather data of Narni, it is possible to individuate the hottest calendar week as the one from July 6th to 12th. And, by analyzing the maximum air temperatures, is possible to select the hottest day of such a week (July 7th—maximum dry-bulb air temperature of 37.5 °C) and adjust accordingly the period of analysis with ± 3 calendar days (i.e., from July the 4th to the 10th). And, as mentioned in Chap. 3, Sect. 3.2, analysing only from 11:00 to 16:00. Hence, for environmental modelling, weather data from the .EPW for July the 4th to the 10th between 11:00 and 16:00 was used. Meanwhile, for air pollutant dispersion modelling, the same timeframe was used to set wind velocity and wind direction (i.e., blowing at 2.4 m/s and 290° from north) and the weather data from the .EPW (on the rest of the climatic parameters) for the hottest day (July 7th) from 11:00 to 16:00. As mentioned in Chap. 1, Sect. 1.2.1, materiality plays an important role in heat stress. The radio-optic properties of the BE surface properties, in particular the surface albedo, describe the proportion of incident radiation reflected by a system. Thus, the amount of heat bounced from the surroundings and the effect that this can have on wind flows (i.e., convective air movement—stack ventilation). Likewise, the roughness of the surface. Such properties are available on material databases from building design standards or literature (e.g., ASHRAE 90.1 [13]), or in the case of the albedo, can be obtained from on-site measurements using certified albedometers, and/or from laboratory measurements with spectrophotometers. Hence, based on in-situ material surveying, averaged façade, roof and ground surface albedos were assigned to the BET2 surfaces as 0.57, 0.51 and 0.2 respectively, with no roughness. For the real-world case scenario, ground surface albedo and all surfaces roughness were equally considered, while 4 different combinations of roof and façade albedo values where set as follows (see Fig. 4.1):

- A—roof 0.5, facade 0.7;
- B—roof 0.5, facade 0.6;
- C—roof 0.5, facade 0.5; and,
- D—roof 0.6, façade 0.6.

---

[9] https://www.ashrae.org/ (last access: 23/10/2023).

4.1 Typological and Real World Scenarios for the Application ...

Such properties are common values for: roof finishings of aged red ceramic tiles (0.5) and clear ceramic tiles (0.6); façade finishings of yellow plaster (0.5), white plaster (0.7) and light stone and travertine (0.6); and grey "sanpietrino" pavement stone (0.2). While the elements density and heat capacity were considered for construction assembly databases (i.e., ASHRAE 90.1 [13] or CBECS[10]) embedded in the default construction set library of Ladybug Tools. For the finishing materials of these construction assemblies, the allocated density and heat capacity values are respectively: 2530 kg/m^3 and 855 J/(kg K) for the rock paved floor; 1920 kg/m^3 and 790 J/(kg K) for the tiled roof; and, 1858 kg/m^3 and 836 J/(kg K) for the rendering (stucco) finishing of walls.

As displayed in Fig. 4.1, no relevant vegetation is found present in the public open space. Thus, no information is gathered on the possibility of existing vegetation proving shade or evapotranspiration for cooling down. Neither for air pollutants deposition. Background air pollution conditions are determined as described in Chap. 3, Sect. 3.2. Air pollutant concentration near the site of interest can be downloaded from regional (ARPA Umbria[11]) or European data repositories (EEA—European Environmental Agency[12]). As done for climatic data, the same pollution data is used in both the BET and the recreated scenario. The median values of the monitored air pollutant concentrations registered from July the 4th to the 10th, between 11:00 and 16:00 were found and set to be, as background pollutant concentration, 4, 122, 23 and 14 μg/m^3 for NO_2, O_3, PM10 and PM2.5 respectively. Pollutant sources different from traffic are excluded as authors have no certainty of the plants or fumes produced at the location, and these were not found during the site survey. Traffic can be set by performing detailed traffic monitoring, or from existing databases. In this case, it is set from the default settings of ENVIMET, referring to the carriageways shown in Fig. 4.1 (considered as two lanes urban roads), with a medium high intensity of 8000 vehicle every 24H, composed of:

- Light duty vehicle 5% (400);
- Heavy duty vehicles 2.5% (200);
- Motorcycles 0.5% (40);
- Public transportation bus 3% (240);
- Coaches 1% (80); and,
- Cars 88% (7040).

---

[10] https://www.eia.gov/consumption/commercial/building-type-definitions.php (last access: 28/10/2023).
[11] https://www.arpa.umbria.it/ (last access: 28/10/2023).
[12] https://www.eea.europa.eu/en (last access: 28/10/2023).

A summary of the data obtained and used for the simulation in both case studies is presented in Table 4.2.

Table 4.2 Summary of the set input data to estimate and map the SLODs hazard risk on the typological and real-case scenarios

Input type	Parameter	Value
Morphological (geometrical)	x–y projected shape	See Fig. 4.1. Resulting virtual models top view of (a) BET2 and (b) Piazza dei Priori in Narni. Measurements are all in meters. The north is upwards, and the square ground level is flat (Fig. 4.1)
	z location and height	
	Geolocation	Narni, Italy (N42.51, E12.51)
	Orientation	332° from north
	Grid analysis shape	Quadratic and triangular
	Grid analysis z location	0.9 from the ground
	Grid size and spacing	174 (BET2A) and 218 (Piazza dei Priori) points every 5 m max
Climatic	Climate class	Cfa
	Dry-bulb air temperature	34.35 °C (50 percentile)
	Relative humidity	33% (50 percentile)
	Atmospheric pressure	96,165.5 Pa (50 percentile)
	Solar irradiation (direct, diffuse, and total)	612, 256, 718 Wh/m^2 (50 percentile)
	Wind speed and direction	3.6 m/s and 250° from north (50 percentile)
	Prevalent wind speed and direction	2.4 m/s and 290° from north
	Hottest week	July 4th to the 10th
	Hottest day	July 7th
Materiality	Surface albedo (reflectivity)	Area weighted average of 0.57 for façade, 0.51 for roof and 0.2 for ground
	Surface transmittance	0.5 W/(m^2 K) for roof, 1.034 W/(m^2 K) for walls
	Density	2530 kg/m^3 for floor; 1920 kg/m^3 for roof; and, 1858 kg/m^3 for wall finishings

(continued)

4.2 Time-Dependent Assessment of Users' Vulnerability and Exposure 97

Table 4.2 (continued)

Input type	Parameter	Value
	Heat capacity	855 J/(kg K) for floor; 790 J/(kg K) for roof; and 836 J/(kg K) for wall finishings
Pollution	Background pollution	$NO_2 = 4$, $O_3 = 122$, PM10 = 23 and PM2.5 = 14 $\mu g/m^3$
	Pollutants sources	Traffic only
	Pollutant emission rate profiles	Constant

4.2 Time-Dependent Assessment of Users' Vulnerability and Exposure

The assessment of users' factor dynamics and of the related KPIs is performed according to the methodology described in Chap. 2, Sect. 2.4. To this end, Fig. 4.2 traces the main areas of the BET (Fig. 4.2a) and of the Narni case study (Fig. 4.2b) by distinguishing their intended use, according to the classification of Chap. 2, Sect. 2.4.2, Table 2.5. In particular, BET data refer to typological, that is recurring, users' exposure and vulnerability, and they have been derived by defining median conditions of the related indicators, as developed in previous works for the historical Italian context [6]. In this sense, homogeneous buildings intended uses have been considered in this section assessment. On the contrary, data for Piazza dei Priori have been retrieved through remote surveys via online mapping tools (e.g., Google Maps), thus ensuring the association between the effective intended use per buildings and at the different stories. The same sources have been also used to determine the opening times of activities in the Narni case study.

Results on Chap. 2 KPIs are then summarized in Fig. 4.3 using the main visualization by time-dependent values, for both working days (W) and holidays (H), and they can point out main differences among the typological and real-world scenario conditions. First, the case of Piazza dei Priori seems to be characterized by higher exposure in respect of those of the reference BET2, as shown by the overall users' outdoor density (UOd_t) (pp/m^2) (Fig. 4.3a). This result is affected by the presence of larger public buildings, which increase the potential number of users, especially during their opening hours (i.e., morning and afternoon). In this sense, both the BET2 and the piazza share the same trend of users' normalized number (NUnt) (Fig. 4.3b), including the exposure peak arousal, to the same time intervals, i.e. the hottest hours of the day (11:00–16:00). Finally, during holidays, the exposure increases again during evening, in view of the attractiveness due to recreational areas, such as bars and restaurants.

The percentage of users for familiarity (FUUrt) (Fig. 4.3c) demonstrates that the Piazza dei Priori is generally less vulnerable than the BET2, since the time-dependent values are placed lower in respect to the typological-related ones, in both holidays and working days. Nevertheless, in working days, during the hottest time of the day, values

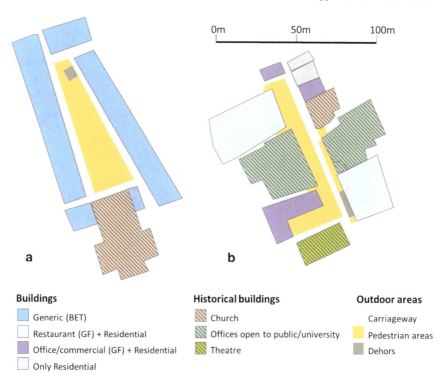

Fig. 4.2 Intended uses of indoor and outdoor areas considered for the BET (**a**) and in the real-world case study of Piazza dei Priori in Narni (**b**). Specific intended uses are classified according to Chap. 2, Table 2.5 data on occupant density, while generic (BET) use refers to a median users' density of 0.2 pp/m² for working days and 0.12 pp/m² for holidays [6]

are close to 0.9 and closer to the ones of the BET2, by conservatively considering that most of users are unfamiliar with the BE and its risks. A similar trend is shared by the ratio between outdoor and indoor users (OUIrt) (Fig. 4.3d), which, for the case study, is lower than expected for the whole typological scenarios. Nevertheless, in this case, it can be argued that only a limited number of users (essentially, less than 20%) seems to be directly exposed to SLOD risks outdoors during the hottest day time depending on their position. The same result is also remarked by the only outdoor (OOp) and prevalent outdoor (POp) percentages (median and max values) shown in Fig. 4.3g. Nevertheless, the OUIrt values are still higher and more constant during the holidays, in view of the opening of most of recreational areas outdoors, including dehors, and of the presence of possible visitors in urban sights. Finally, the percentage of users for vulnerability due to age and ideal lack of autonomy in the built environment use (VUrt) (Fig. 4.3e) and the overall vulnerability-by-age index (Vat) (Fig. 4.3f) show the same (but vertically translated) trend, since most of the users' typologies depending on age-class are equally distributed over time. The same trend is noticed in both BET2 and case study scenarios, although Narni's data are

4.2 Time-Dependent Assessment of Users' Vulnerability and Exposure

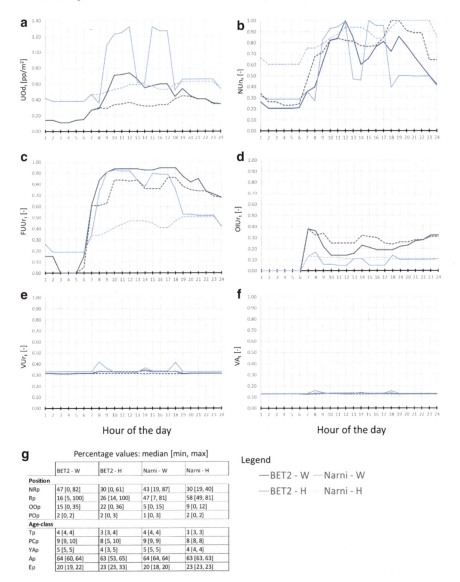

Fig. 4.3 Results on the BET2 and the case study of Narni, for working days (W) and holidays (H) concerning the KPIs on users' exposure and vulnerability quick characterization, as defined in Chap. 2, Sect. 2.4.4. X-axes are expressed in hours of the day, and time-dependent values concern the overall users' outdoor density (panel A), the users' normalized number (panel B), the percentage of users for familiarity (panel C), the ratio between outdoor and indoor users (panel D), the percentage of users for vulnerability due to age and ideal lack of autonomy in the built environment use (panel E), the overall vulnerability-by-age index (panel F). Panel G traces basic statistics for percentage of users' typology in terms of position and age-class

over the typological ones for working days. This points out that the case study can be characterized by an ideal higher individual vulnerability than other similar scenarios (while relating to median values). These results are also remarked by the percentage values on age-class typologies, which shows that most of users (more than 60%) in the BET2 and in the case study are adult users (Fig. 4.3g), for both holidays and working days.

Finally, KPIs about users' relevant health frailty data have not been addressed since no specific databases have been directly and freely found for the whole sample and for the case study.

In view of the above, and according to related works on the BET definition and applications to risk assessment [5, 7, 14], we can conclude that:

1. Typological outcomes on users' factors are quite close to those of the case study, thus suggesting that a very preliminary assessment on the BET can delineate general issues about users' factors, and that the case study could be representative of a wider sample too;
2. The users' exposure for the case study is greater than the one of the typological conditions, while the users' vulnerability due to their position in the BE is lower in the case study than in the typological conditions. These issues remark the importance of defining specific analysis depending on the effective scenarios, which can tailor the assessment and mitigation tasks;
3. The users' vulnerability due to age-classes is the same for both typological and case study scenarios, thus remarking the importance of previous point 1.

4.3 "Behavioural-Based" Assessment of SLODs Risk in the Current Scenario

This section adopts the operational flowchart for "behavioural-based" approach described in Chap. 3, Sect. 3.1. In particular, the SLOD simulation tools to predict the UTCI and AQI values in the open spaces are applied depending on the scenario creation data about the BE and climate features (defined in Sect. 4.1), and then, the effects on users' behaviours due to the UTCI and outdoor use are calculated (Sect. 4.3.1).The behavioural-based simulation results on UTCI (Sect. 4.3.2) and air pollution (Sect. 4.3.3) effects on users' health are assessed through the proposed KPIs, to characterize the current risk level for the scenarios. In the following subsections, the results of each methodology step are provided contemporarily for the BET and the real-world case study of Narni. Finally, the multirisk assessment (Sect. 4.3.4) between the two case studies have been performed according to Chap. 3, Sect. 3.4, to additional point out which scenario can involve worst conditions to SLODs.

4.3.1 From SLOD Simulation to the Assessment of Users' Distribution in the Public Open Spaces

The SLOD simulation-based assessment of the "behavioural-based" approach is performed through the methods described in Chap. 3, Sect. 3.1, after having defined and prepared the scenarios in Sect. 4.1. The derived micro-climatic conditions of heat stress and air quality distress in the form of the Universal Thermal Climate Index (UTCI) and Air Quality Index (AQI) are mapped within the public open space in both BET2 and Piazza dei Priori (see Fig. 4.4).

The results obtained for heat stress mapping from BET2 resemble the results seen in Piazza dei Priori (see Fig. 4.4a, b). A lower heat stress is noticed in the narrower area of the public open space, while on the wider part, higher UTCI results are present. In addition, given its orientation, the western buildings provide shade into the left side of the open space, decreasing the perceived heat stress, as a consequence of the physical vulnerability factors, as highlighted in Chap. 1, Sect. 1.2.1. UTCI values are ranging from 33 to 38 °C (or above), that is having heat stress categories of strong to very strong (see Table 3.1 in Chap. 3). Heat stress conditions that should be avoided over long periods of time. Results are then discussed, in Sect. 4.3.2.

While mapping AQI, volumes and analysis grid presented a shift (see Fig. 4.5c, d) due to the misalignments between the geometric modelling capabilities of ENVIMET v5 and Rhinoceros v6. In fact, ENVIMET v5 allows quadratic elements only. Moreover, air pollution distress distribution differs between BET2 and Piazza dei Priori. However, these variations are probably due to the different traffic settings, and the sensitivity of Computational Fluid Dynamics (CFD) to sharp edges and narrow spaces that modify wind velocity [15, 16] (as described in Chap. 1, Sect. 1.2.1). As a consequence, lower wind velocity at the corners implies lower air pollutant dispersion [17, 18]. Nevertheless, the fact that the air pollution distress on the northeast side of the public open space is higher than the rest of the area prevails in both cases. It should be noted that air distress, according to air quality categories, is low (good air quality as AQI < 50) in both scenarios. Despite of having such category, results should be handle with care as no other pollutant sources have been included; thus, the risk of the air pollution distress is assessed based on an idealized condition (see Sect. 4.3.3).

For the 3rd phase of the proposed "behavioural" approach for SLOD risk presented in Chap. 3, Sect. 3.1, the potential distribution of people within the open public space is estimated based on their response to heat stimulus (Chap. 2, Sect. 2.3, and Chap. 3, Sect. 3.3.1). From the resulting UTCI—heat stress distribution, the acceptability of the micro-climate conditions for prevalent outdoor (PO) and only outdoor (OO) users is mapped in BET2 and Piazza dei Priori and presented in Fig. 4.5 (using Eqs. 3.1 and 3.2).

The thermal acceptability, and thus, the potential users' distribution follow the inverse of the UTCI trend. Occupation density is hence expected to be higher where lower UTCI values are estimated, and both scenarios are similar to each other (Fig. 4.4). Thus, people will can gather next to buildings shade, where they are less vulnerable to heat stress and air quality distress (Chap. 1, Sect. 1.2.1). Although,

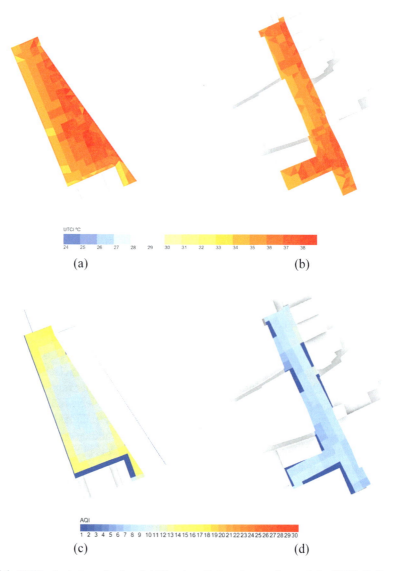

Fig. 4.4 UTCI—heat stress (top) and AQI—air pollution distress (bottom) for BET2 (left) and Piazza dei Priori (right). Results correspond to the averaged conditions found between 11:00 and 16:00 during July 7th for AQI, and from July 4th to 10th for UTCI

a relevant occupation density on the southwest for both case studies might influence the risk of people exposed to lower quality of air (higher air pollutant concentrations). These conditions are worse for OO users (see Fig. 4.4c, d), for which their acceptability ranges between 85 to nearly 100%, compared to PO which is always below 75%.

4.3 "Behavioural-Based" Assessment of SLODs Risk in the Current Scenario 103

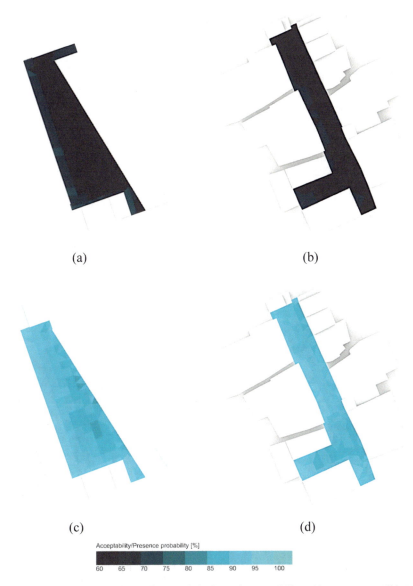

Fig. 4.5 Distribution of users depending on their thermal acceptability of heat stress conditions for PO (top) and OO (bottom), for BET2 (left) and Piazza dei Priori (right). Results correspond to the computed acceptability using Eqs. 3.1 and 3.2 from UTCI averaged conditions found between 11:00 and 16:00 from July 4th to 10th

4.3.2 Heat Stress Assessment and Effects on Health

Based on the results obtained from Sect. 4.3.1, and following the procedure mentioned in Chap. 3, Sect. 3.3.3, the last part of the 3rd phase of the proposed "behavioural-based" approach for SLOD risk assessment is carried out. Estimating the heat related KPI on health effect, based on potential water loss through sweat, the likelihood of risk was quantified (WLR$_{group}$—water loss on body weight per group rate risk, Chap. 3, Eq. 3.5) and summarized in Table 4.3 for PO and Table 4.4 for OO public open space users.

Results in Table 4.3 and Fig. 4.4 validate the resemblance of the use of BETs for analyzing this type of SLOD risk. In fact, the average of the arithmetical difference between the WLR in the typological scenario and the real-world scenario is < 0.006%. Moreover, the obtained results are aligned with what has been presented by

Table 4.3 Percentage (%) of water loss on body weight (WLR$_{group}$) given the potential 1 h exposure of the users performing 1 h behavior (PO), and perceived thermal stress within BET2 and Piazza dei Priori, given a certain age class. Codes relating to age groups, defined in Chap. 3, Sect. 3.3.2, are: toddlers (TU), children (PC), young adults (YA), adults (AU), elderly (EU)

Gender and age class	BET2	Piazza dei Priori
Male/female TU	1.65	1.67
Male/female PC	0.45	0.46
Male YA	0.24	0.24
Female YA	0.28	0.28
Male AU	0.2	0.21
Female AU	0.24	0.24
Male EU	0.22	0.22
Female EU	0.26	0.26

Table 4.4 Percentage (%) of water loss on body weight (WLR$_{group}$) given the potential 1 h exposure of the users performing transient behavior (OO), and perceived thermal stress within BET2 and Piazza dei Priori, given a certain age class. Codes relating to age groups, defined in Chap. 3, Sect. 3.3.2, are: toddlers (TU), children (PC), young adults (YA), adults (AU), elderly (EU)

Gender and age class	BET2	Piazza dei Priori
Male/female TU	2.47	2.50
Male/female PC	0.68	0.69
Male YA	0.35	0.36
Female YA	0.42	0.42
Male AU	0.31	0.31
Female AU	0.36	0.36
Male EU	0.33	0.33
Female EU	0.39	0.39

Blanco Cadena et al. [5], in which toddlers are at a higher risk compared to elders (approximately ×6), the type of behavioral response to heat stress can generate + 1/3 of risk (i.e., PO vs. OO), and that a toddler exposed for 2 h or more (~ 4% or above water loss on body weight, either for OO or PO) can reach a dehydration risk state and related pathologies [19].

4.3.3 Air Quality Assessment and Effects on Health Due to Pollution

The same procedure is carried out for air quality risk estimates. Calculating instead short-term pollution risk (STPR) according to the type of health affection, based on pollutant concentration relative increase. Calculations are performed with Eqs. 3.7, 3.8, 3.9 and 3.10 concentrating only on NO_2, given that these are more sensitive to traffic pollution emissions [20]. Hospital admissions with cardiovascular issues and mortality are the only health affections treated, due to the lack of relative risk databases (see Chap. 3, Sect. 3.3.3). Results are shown in Tables 4.5 and 4.6.

In the case of air pollution SLOD risk assessment the BET2 did not perform well compared to the real case study as mentioned earlier, due to the differences in traffic settings. Nevertheless, results remain valid as they are able to compute the augmented risk of being in an open space with such characteristics. That is the case for Piazza dei Priori, in which the type of behaviour can determine an approximate increase between ×1 1/3 and ×2. Nevertheless, the risk of either mortality or cardiovascular remains low. Only under the conditions of BET2, for 2 h an increase in mortality could reach 1% by being exposed for 2 h having an OO user behaviour.

Table 4.5 Short term pollution risk ($STPR_{NO_2}$) (%) given the potential 1 h exposure to NO_2 of the users performing 1 h behaviour (PO) within the public open space in BET2 and Piazza dei Priori, compared to a control group in an air-pollutant-free environment

Affections	BET2	Piazza dei Priori
Hospital admission with cardiovascular issues	0.21	0.03
Mortality	0.38	0.05

Table 4.6 Short term pollution risk ($STPR_{NO_2}$) (%) given the potential 1 h exposure to NO_2 of the users performing transient behaviour (OO) within the public open space in BET2 and Piazza dei Priori, compared to a control group in an air-pollutant-free environment

Affections	BET2	Piazza dei Priori
Hospital admission with cardiovascular issues	0.30	0.06
Mortality	0.54	0.11

Table 4.7 Multirisk assessment for BET2 and the Narni case study through the multirisk metrics MR and its components proposed in Chap. 3, Sect. 3.4

Parameter	BET2	Piazza dei Priori
Peak UOd_t (pp/m^2)	0.75	1.3
EV (%)	0.75	1.30
HS (–)	0.515	0.521
AP (–)	0.007	0.001
MR (%)	0.196	0.339

4.3.4 Multirisk Assessment

The multirisk assessment adopts the approach shown in Chap. 3, Sect. 3.4, Eq. 3.13, which allows to combine: (1) the impact of the users' exposure and vulnerability, focusing on the users' density and the presence of toddlers; (2) the related impact of heat stress on the toddlers; (3) the related impact of air pollution distress on the toddlers. Table 4.7 summarized the calculation process and the final results according to the proposed methodology. The results are calculated on peak conditions for crowding in the scenario (see Sect. 4.2) within the assessed time span (11:00–16:00), according to a conservative approach. Furthermore, in both cases, TU% = 4%. *MR* points out the main differences discussed above between the idealized case study of the BET and the real-world scenarios. In particular, the Piazza dei Priori scenario is riskier than the BET2 in view of the main impact of UOd_t and of the (although slight) higher *HS* conditions.

4.4 Risk Mitigation Strategies Evaluation

The mitigation strategies outlined in Chap. 3, Sect. 3.4, have been proposed and tested according to the "behavioural-based" approach (compare with Chap. 3, Sect. 3.1, Fig. 3.1). In particular, the selected scenario is the one related to the case study that suffers from higher multirisk conditions, so as to better stress the impact of mitigation strategies on the more critical examined conditions. As discussed in Sect. 4.3.4, Table 4.7, the real-world scenario of Piazza dei Priori displays a higher UOd_t and more significant *HS* for TU, which lead to higher level of multirisk according to *MR* values. The *MR* assessment is applied at the end of this section to evaluate differences from a multirisk perspective, once mitigation strategies have been applied and related simulations have been performed.

4.4 Risk Mitigation Strategies Evaluation

Figure 4.6 hence resumes the selected risk mitigation strategies for the application in the real-world case study, which have been adapted depending on Piazza dei Priori peculiarities in terms of historical context and intended use of indoor and outdoor areas. In particular, the strategies have been selected in view of their applicability, their potential beneficial effects on the risk levels and on the users (in terms of health impacts and behaviours in outdoor areas use) as discussed in Chap. 3, Sect. 3.4, Table 3.6. Moreover, specific solutions have also to ensure their compatibility with the building heritage identity and preserve main features. Thus, for instance, specific solutions which can consistently vary the aesthetic of building façade (e.g., green walls), introducing "foreign" elements in the outdoor spaces (e.g., large water bodies, trees or shrubs and edges), or modifying the materiality of outdoor surfaces of public spaces (e.g., green pavements), have been excluded. On the contrary, addition of components and obstacles in the public open spaces are considered (e.g., see A3, A4), although they can partially modify the outdoor areas use. Although having a limited surface application, the impact of such strategies has been deeply assessed through simulation since they can provide shade and obstruct the wind flow, thus affecting both the SLODs.

In view of the above, Fig. 4.6 also traces the main implementation notes, which can support the deployment of the solution within the BE, as well as gives a visualization of the implementation of solutions by pointing out the reference application elements (i.e., the buildings) and the position of outdoor components (e.g., those comprising urban furniture). Additional details on the retrofit scenarios are also summarized in Table 4.8.

Thanks to the simulation-based methodology, as done for the as-built conditions, health impacts of heat stress (i.e., WLR) are estimated for each mitigation strategy. These are presented and summarized in Tables 4.9 and 4.10, then compared with the baseline to determine the most suitable solution.

From the results obtained, A3, B5 and B6 have no effect on heat-stress related risk. In particular, A3 has no specific impact in view of no direct shadings effects of the implemented greenery. Nevertheless, A3 can have a higher impact on the movement of the users being obstacles to their free use of outdoor spaces (compare with Chap. 3, Sect. 3.4). B5 and B6 provoke a generalized effect on air temperature that is more effective on the macro-scale, rather than at the scale of the square, essentially because of low level of homogeneity in their application on the buildings [21]. A4 remains a very spot-on mitigation strategy that brings little or no effect for the whole public open space. Such a result seems also to be essentially correlated to the implementation position, on the west side of the square, in a part of the outdoor area which could be partially shaded by the buildings indeed. As a consequence, it is worth noticing that the balance between specific possible position and extent of interventions (which depends on the BE layout and use) and their positive impacts should be accurately evaluated in the design phase. For PO users, the most suitable solution to employ would be to implement B1 or B2, according to simulation results and KPIs. Indeed, cool pavement mitigation strategy has been demonstrated to, in overall, lower down the effect of urban heat island in urban BEs [21]. Nevertheless, this solution should be studied in depth, and a more wider scale on the urban fabric,

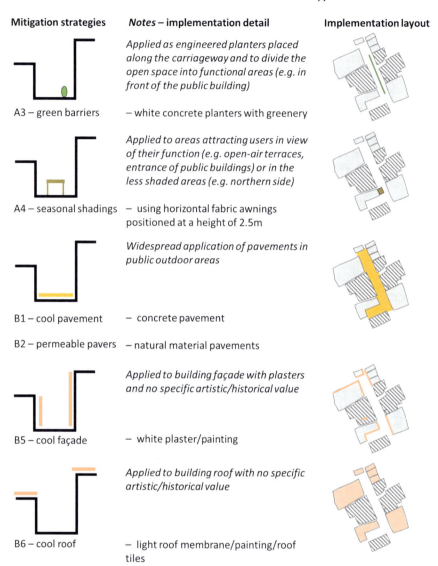

Fig. 4.6 Overview of selected mitigation strategies and their implementation within and the real-world case study. In the implementation layout, building open to the public with historical values are marked by the hatch

as higher albedo increases the amount of radiation perceived by the public open space users, thus increasing the heat stress (i.e., UTCI), and in consequence making those spaces less utilized. In fact, looking at the UTCI, and acceptability mapping of B1 and B2, the UTCI levels are lower for B2 (ranging from 34 to 38 °C) than for B1 (from 33 to 43 °C) (see Fig. 4.7), but the acceptability of PO users has driven them

4.4 Risk Mitigation Strategies Evaluation

Table 4.8 Summary of the input data required for the mitigation strategies to estimate and map the SLODs hazard risk on the real-world case scenarios and compare their efficacy with the baseline

Input type	Parameter	Value
Materiality	Surface albedo (reflectivity)	Cool roof 0.8 (B6), cool façade 0.8 (B5), Cool pavement 0.8 (B1), permeable pavement 0.08 (with soil saturation at 0.395—B2), greenery 0.2 (A3), fabric awnings 0.65 (A4)
	Surface transmissivity	Fabric awnings 0.25 (A4), greenery 0.3 (A3)
Vegetation	Leaf area index	0.3 (A3)
	Height	Planters of 1 m height with plants of 0.15 m height (A3)

Table 4.9 Percentage (%) of water loss on body weight (WLR$_{group}$) given the potential 1 h exposure of the users performing 1 h behavior (PO), and perceived thermal stress within Piazza dei Priori depending on the mitigation strategy, given a certain age class. Codes relating to age groups, defined in Chap. 3, Sect. 3.3.2, are: toddlers (TU), children (PC), young adults (YA), adults (AU), elderly (EU). In addition, values into brackets express the dKPIs,r percentage variation for the strategy in respect to the baseline values as reference condition (see Table 4.3)

Gender and age class	A3	A4	B1	B2	B5	B6
Male/female TU	1.67 (0%)	1.67 (0%)	1.54 (− 8%)	1.69 (1%)	1.67 (0%)	1.67 (0%)
Male/female PC	0.46 (0%)	0.46 (0%)	0.42 (− 9%)	0.47 (2%)	0.46 (0%)	0.46 (0%)
Male YA	0.24 (0%)	0.24 (0%)	0.22 (− 8%)	0.24 (0%)	0.24 (0%)	0.24 (0%)
Female YA	0.28 (0%)	0.28 (0%)	0.26 (− 7%)	0.29 (4%)	0.28 (0%)	0.28 (0%)
Male AU	0.21 (0%)	0.21 (0%)	0.19 (− 10%)	0.21 (0%)	0.21 (0%)	0.21 (0%)
Female AU	0.24 (0%)	0.24 (0%)	0.22 (− 8%)	0.24 (0%)	0.24 (0%)	0.24 (0%)
Male EU	0.22 (0%)	0.22 (0%)	0.2 (− 9%)	0.22 (0%)	0.22 (0%)	0.22 (0%)
Female EU	0.26 (0%)	0.26 (0%)	0.24 (− 8%)	0.27 (4%)	0.26 (0%)	0.26 (0%)

Table 4.10 Percentage (%) of water loss on body weight (WLR$_{group}$) given the potential 1 h exposure of the users performing transient behavior (OO), and perceived thermal stress within Piazza dei Priori depending on the mitigation strategy, given a certain age class. Codes relating to age groups, defined in Chap. 3, Sect. 3.3.2, are: toddlers (TU), children (PC), young adults (YA), adults (AU), elderly (EU). In addition, values into brackets express the dKPIs,r percentage variation for the strategy in respect to the baseline values as reference condition (see Table 4.4)

Gender and age class	A3	A4	B1	B2	B5	B6
Male/female TU	2.5 (0%)	2.5 (0%)	2.91 (16%)	2.64 (6%)	2.52 (1%)	2.51 (0%)
Male/female PC	0.69 (0%)	0.69 (0%)	0.8 (16%)	0.73 (6%)	0.69 (0%)	0.69 (0%)
Male YA	0.36 (0%)	0.36 (0%)	0.42 (17%)	0.38 (6%)	0.36 (0%)	0.36 (0%)
Female YA	0.42 (0%)	0.42 (0%)	0.49 (17%)	0.45 (7%)	0.43 (2%)	0.42 (0%)
Male AU	0.31 (0%)	0.31 (0%)	0.36 (16%)	0.33 (6%)	0.31 (0%)	0.31 (0%)
Female AU	0.36 (0%)	0.36 (0%)	0.42 (17%)	0.38 (6%)	0.36 (0%)	0.36 (0%)
Male EU	0.33 (0%)	0.33 (0%)	0.39 (18%)	0.35 (6%)	0.33 (0%)	0.33 (0%)
Female EU	0.39 (0%)	0.39 (0%)	0.46 (18%)	0.42 (8%)	0.4 (3%)	0.39 (0%)

out of the public open space in B1 (minimum acceptability is approximately 10%). This is clearer while looking at the risk conditions for OO users for B1, as it is the highest. These results reinforce the need for a behavioural approach to be considered while studying SLOD risks, and their potential effect for preparing the public open space occupation density for sudden onset disasters (e.g., earthquakes).

Therefore, from this analysis and from a heat stress perspective, a combination between B1 and B2, which has a higher albedo of the surface in B2 but retains the benefits of evaporative cooling, could serve as a proper solution for the public open space. Hence, such an integrated approach could promote the usability of the space and the health of its users. Instead, the effects of the mitigation strategies on the NO_2 concentration distribution and air pollution distress ($STPR_{NO_2}$) for the users of Piazza dei Priori are gathered in Table 4.11 for PO and Table 4.12 for OO.

The obtained results suggest that the proposed mitigation strategies have no or little effect on the actual air pollutant risk for public open space users in Piazza dei Priori. However, considering individual vulnerability, the typology of user, and their implied behaviour, suggests that their risk of both hospital admission with cardiovascular issues and mortality could increase $\times 2$ if they are more prone to tolerate higher heat stress under the resulting microclimatic conditions. In view of these outcomes, it could be noticed that the strategies do not seem to significantly modify wind patterns, and thus the distribution of air pollution concentrations. Nevertheless, this outcome seems to be in contrast with previous work suggestions [22], but it could be due to the taller greenery (up to 3 m) considered in previous analysis, because such a shape can effectively represent a greater barrier to wind flow. Finally, multirisk assessment is performed as for Sect. 4.3.4, for each of the strategy, and results are shown in Table 4.13. MR demonstrates what is reported separately for heat stress and air pollution effects, thus underlining that strategies are not generally effective in relation to the toddlers since the mitigation impact is very limited. However, some strategies (i.e. B1 and B2) seems to be risk-increasing in view of their effects on the toddlers, especially for the HS values. This outcome is a consequence of the combination of OO and PO effects in the final parameters, and thus the overall view should be compared to the specific outcomes of the specific risk-affecting conditions, as discussed above.

4.4 Risk Mitigation Strategies Evaluation

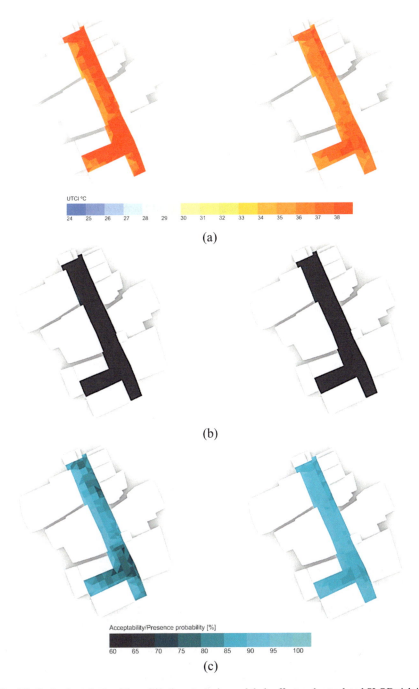

Fig. 4.7 In depth analysis of the mitigation strategies and their effect on heat related SLOD risk in Piazza dei Priori. Displaying for B1 (left) and B2 (right) the results on **a** UTCI, **b** acceptance for PO users, and **c** acceptance for OO users

Table 4.11 Short term pollution risk (%) given the potential 1 h exposure of the users performing transient behaviour (PO) to NO_2 within the public open space in Piazza dei Priori depending on the mitigation strategy, compared to a control group in an air-pollutant-free environment

Affections	A3	A4	B1	B2	B5	B6
Hospital admission with cardiovascular issues	0.03 (0%)	0.03 (0%)	0.03 (0%)	0.03 (0%)	0.03 (0%)	0.03 (0%)
Mortality	0.05 (0%)	0.05 (0%)	0.05 (0%)	0.05 (0%)	0.05 (0%)	0.05 (0%)

In addition, values into brackets express the dKPIs,r percentage variation for the strategy in respect to the baseline values as reference condition (see Table 4.5)

Table 4.12 Short term pollution risk (%) given the potential 1 h exposure of the users performing transient behaviour (OO) to NO_2 within the public open space in Piazza dei Priori depending on the mitigation strategy, compared to a control group in an air-pollutant-free environment

Affections	A3	A4	B1	B2	B5	B6
Hospital admission with cardiovascular issues	0.06 (0%)	0.06 (0%)	0.06 (0%)	0.06 (0%)	0.06 (0%)	0.06 (0%)
Mortality	0.11 (0%)	0.11 (0%)	0.11 (0%)	0.11 (0%)	0.11 (0%)	0.11 (0%)

In addition, values into brackets express the dKPIs,r percentage variation for the strategy in respect to the baseline values as reference condition (see Table 4.6)

Table 4.13 Multirisk assessment for the Narni case study through the multirisk metrics MR and its components proposed in Chap. 3, Sect. 3.4, comparing pre-retrofit (pre) and mitigation strategies scenarios

Parameter	Pre	A3	A4	B1	B2	B5	B6
HS (%)	0.521	0.521	0.521	0.556	0.541	0.524	0.523
AP (%)	0.001	0.001	0.001	0.001	0.001	0.001	0.001
MR (%)	0.339	0.339	0.339	0.362	0.352	0.341	0.341

References

1. Gherri B et al (2021) On the thermal resilience of Venetian open spaces. Heritage 4:4286–4303. https://doi.org/10.3390/heritage4040236
2. Rosso F et al (2018) On the impact of innovative materials on outdoor thermal comfort of pedestrians in historical urban canyons. Renew Energy 118:825–839. https://doi.org/10.1016/j.renene.2017.11.074
3. Speak AF, Salbitano F (2022) Summer thermal comfort of pedestrians in diverse urban settings: a mobile study. Build Environ 208:108600. https://doi.org/10.1016/j.buildenv.2021.108600
4. Kubilay A et al (2019) Coupled numerical simulations of cooling potential due to evaporation in a street canyon and an urban public square. J Phys Conf Ser 1343:012016. https://doi.org/10.1088/1742-6596/1343/1/012016
5. Blanco Cadena JD et al (2023) Determining behavioural-based risk to SLODs of urban public open spaces: key performance indicators definition and application on established built environment typological scenarios. Sustain Cities Soc 95:104580. https://doi.org/10.1016/j.scs.2023.104580

References

6. Quagliarini E (2022) Users' vulnerability and exposure in public open spaces (squares): a novel way for accounting them in multi-risk scenarios. Cities 133:104160
7. D'Amico A et al (2021) Built environment typologies prone to risk: a cluster analysis of open spaces in Italian cities. Sustainability 13:9457. https://doi.org/10.3390/su13169457
8. You W, Ding W (2021) Effects of urban square entry layouts on spatial ventilation under different surrounding building conditions. Build Simul 14:377–390. https://doi.org/10.1007/s12273-020-0656-8
9. da Silva FT et al (2022) Influence of urban form on air quality: the combined effect of block typology and urban planning indices on city breathability. Sci Total Environ 814:152670. https://doi.org/10.1016/j.scitotenv.2021.152670
10. Morganti M (2021) Spatial metrics to investigate the impact of urban form on microclimate and building energy performance: an essential overview. In: Palme M, Salvati A (eds) Urban microclimate modelling for comfort and energy studies. Springer, Cham, pp 385–402
11. Belpoliti V et al (2018) A parametric method to assess the energy performance of historical urban settlements. Evaluation of the current energy performance and simulation of retrofit strategies for an Italian case study. J Cult Herit 30:155–167. https://doi.org/10.1016/j.culher.2017.08.009
12. Beck HE et al (2018) Present and future Köppen-Geiger climate classification maps at 1-km resolution. Sci Data 5:1–12. https://doi.org/10.1038/sdata.2018.214
13. Goel S (2017) ANSI/ASHRAE/IES standard 90.1-2016 performance rating method reference manual
14. Quagliarini E et al (2023) How could increasing temperature scenarios alter the risk of terrorist acts in different historical squares? A simulation-based approach in typological Italian squares. Heritage 6:5151–5188. https://doi.org/10.3390/heritage6070274
15. Dupont S, Brunet Y (2008) Edge flow and canopy structure: a large-eddy simulation study. Bound-Layer Meteorol 126:51–71. https://doi.org/10.1007/s10546-007-9216-3
16. Soulhac L et al (2008) Flow in a street canyon for any external wind direction. Bound-Layer Meteorol 126:365–388. https://doi.org/10.1007/s10546-007-9238-x
17. Karttunen S et al (2020) Large-eddy simulation of the optimal street-tree layout for pedestrian-level aerosol particle concentrations—a case study from a city-boulevard. Atmos Environ X 6:100073. https://doi.org/10.1016/j.aeaoa.2020.100073
18. Tiwari A, Kumar P (2020) Integrated dispersion-deposition modelling for air pollutant reduction via green infrastructure at an urban scale. Sci Total Environ 723:138078. https://doi.org/10.1016/j.scitotenv.2020.138078
19. Vanos JK et al (2018) Evaluating the impact of solar radiation on pediatric heat balance within enclosed, hot vehicles. Temperature 5:276–292. https://doi.org/10.1080/23328940.2018.1468205
20. US EPA (2018) Technical assistance document for the reporting of daily air quality—the air quality index (AQI)
21. Santamouris M (2014) Cooling the cities—a review of reflective and green roof mitigation technologies to fight heat island and improve comfort in urban environments. Sol Energy 103:682–703. https://doi.org/10.1016/j.solener.2012.07.003
22. Salvalai G et al (2023) Greenery as a mitigation strategy to urban heat and air pollution: a comparative simulation-based study in a densely built environment. Riv Tema 09. https://doi.org/10.30682/tema090003

Chapter 5
Conclusions and Perspectives

Abstract Risk assessment and mitigation of Slow-Onset Disasters (SLODs) linked to climate-change affecting the urban Built Environment (BE) could be faced through innovative methodologies based on a "behavioural-based" and multi-risk approach. This approach is behavioural-based since relies on the analysis of the users' behaviours, time-dependent BE fruition rules, needs and responses to critical conditions due to several SLODs (e.g., average temperature rise, heat waves, average air pollutant rise, fog). Moreover, the approach is oriented towards multi-risk because it tries to combine the SLODs effects on users' health in different scenarios (in this book, heat stress and air pollution distress) by exploiting simulation models and combined metrics on risk assessment and mitigation strategies effectiveness. In particular, the main application could relate to the urban open spaces, such as streets and squares, where the SLODs effects on the BE users are more evident and linked to behavioural issues outdoors as well as to short-term exposure. This book offers the methodological definition and application to representative case studies of the behavioural-based approach and of its related tools, by then demonstrating the capabilities of the proposed single to multi-risk perspective in the assessment of risk levels and comparison of effectiveness of different mitigation strategies. Nevertheless, the capabilities of the approach use should be extended, and thus further activities are needed. This chapter provides an overview of the main goals reached by the research, as well as of the demonstrated capabilities of the approach. Then, it also traces the next steps for the approach development and application from both the researchers and stakeholders' standpoint (mainly, for designers and local administrations).

Keywords Low-onset disasters · Built environment · Heat · Air pollution · Health · Behavioural-based approach

5.1 Urban Built Environment and Slow-Onset Disasters: How the "Behavioural-Based" Approach Could Support Risk Assessment and Mitigation?

The urban Built Environment (BE) has become the main preferred living location for most of the world population. The United Nations reported in 2018 that 55% of the world's population was living in highly urbanized areas and that it is projected to grow up to 68% by 2050 [1, 2]. Such an issue underlines how cities are the hot spots of exposure to many different risks and disasters. In particular, the World Health Organization [3, 4] foresee that around 6.7 billion people are at great risk due to climate change risks, and thus are associated with critical single and multi-risk conditions in view of related Slow Onset Disasters (SLODs). In addition, the urban BE has demonstrated to have an inherent dynamism, which is more relevant if compared to the ones of rural areas [5, 6], especially considering increasing air pollution (lower CO_2 capture capacity [7]) and temperatures (global average $+0.7–1.7$ °C). Moreover, the spatial components of a urban BE, from its districts down to its neighborhoods, building blocks, and single open spaces (e.g. streets and squares) behave differently, environmentally-wise, according to their characteristics resulting in diverse microclimates [8–13], in view of the physical vulnerability of the urban BE. Considering the physical science of the atmosphere, and the nature of SLODs, it is common that both increasing air pollution and temperature are currently slowly unfolding [6]. Nevertheless, although this process is raising slowly, effects on the communities hosted in our cities (i.e. effects on human health) are significantly emerging over the time due to the combination of hazard and physical vulnerability factors with those related to the dynamics and features of users' exposure and individual vulnerability in the urban BE. In particular, effects on human health, which are predominant in the considered SLODs context, can be more relevant depending on the exposed demographic groups (e.g., based on age and health status—up to $\times 6$ times more, see Chap. 4, Sect. 4.3), allowing to identify and manage social vulnerability as well. Finally, current climate and environmental changes around the globe suggest an increase in air temperature and air pollutants' concentration that will destabilize the fragile balance between anthropogenic activities and nature resilience capacity [14]. In fact, records show $\times 2.7$ times more meteorological disaster events in this century with respect to the last century, and $\times 1.4$ comparing the last decade with the last century [15].

Therefore, the growth of single and multi-risk levels in the urban BE (i.e. intertwined effects of increasing temperatures and air pollution, see Chap. 1, Sect. 1.4) motivates urgent need for a rapid, space and time granular, multi-domain and multi-risk assessment method to equip designers, developers and policy makers for climate change resilience and population safety [6]. This book tries to move in such direction by using a "behavioural-based" approach to SLODs risk assessment and mitigation [16–19], and by also pursuing a multi-risk perspective [17, 18, 20] thanks to the combination of effects on users' health due to both air pollution and heat stress. The "behavioural-based" approach means that the assessment of current SLODs risk in a given urban BE and of effectiveness of mitigation strategies should be evaluated by taking into account users' behaviours and needs in the urban BE fruition and in respect to the environmental stressors. To this end, simulation models jointly represent: (a) the physical vulnerability of the urban BE and related key factors (Sect. 5.1.1); (b) the users' exposure over space and time in respect to their behaviours in the urban BE (Sect. 5.1.2); (c) the users' vulnerability which can additionally later the SLODs effects on human health (Sect. 5.1.3). The multi-risk perspective implies that SLODs effects on users are then combined, by relaying, in the proposed approach, on the users' vulnerability as leading factor for risk assessment (Sect. 5.1.3). All these concepts are mainly oriented to the open spaces in the urban BE as investigated scenarios, since (see Chap. 2) [13, 18, 21, 22]: (1) they are basic components in the urban BE, directly linking indoor areas to the nearest outdoors, and are the spaces where users move and behave daily; (2) they are relevant attractors for users (both citizens and visitors) over time and space, being characterized by relevant dynamics all over the day; (3) they are outdoor areas in which users can directly suffer from environmental conditions and thus stress due to SLODs; (4) their conditions (related to layout, space intended use, and environmental conditions) highly affect the users' attraction to specific areas.

Considering the open space in the urban BE as the main application scenario and scale for the "behavioural-based" approach, the whole methodology aims at a rapid-but-robust support the decision makers, and thus it:

- determines the assessment scale of risk factors according to different coarse networks which relates on features and extent of parameters about: (1) physical vulnerability, i.e. discretization of the outdoor spaces by using consolidated simulation tools and their grids (in this work, 5 × 5 m); (2) background and specific SLODs conditions, i.e. on-site versus standard data collection methods about local climate conditions and environmental stressors; and (3) users' factors, i.e. relationship with the general intended uses of indoor/outdoor areas and with the outdoor spaces discretization grids;
- considers the variability of risk factors over space and time, by considering an hourly discretization of occupancy and microclimatic condition;
- uses evidence-based and context-based data for simulation purposes, by managing them through two levels of analysis, that are: (1) standard scenarios, which ensure application rapidity and robustness since they are based on statistical analysis of large samples of input factors, and the selection of recurring (e.g. median) conditions as representative ones; (2) deeper site-related assessment using monitoring campaigns and surveys, and organizing data, e.g., into time series on environmental and users' parameters;
- reduces the computational burden of yearly climate-based analysis for 8760 h of the year and lengthy air pollution dispersion calculations. In fact, it focuses on the most critical conditions that can be suffered by users in the BE, according to a conservative approach;
- allows to be applied in any other context with open-source databases and tools available to designers, practitioners and policy makers;

- uses simple-but-reliable Key Performance Indicators, such as those based on the final effects of SLODs on users' health due to both short and longer exposure, and their combination (in each SLOD and in the multi-risk perspective); and,
- considers the mitigation strategies alternatives and its design in respect to (1) the specific features of each solution against each of the considered SLODs, (2) the possibility to affect users' behaviours by their application, and (3) the implementation contingency and constraints depending on BE uses and specific features (including heritage-related).

5.1.1 Physical Vulnerability

Urbanization growth demands urban BE to grow and expand (68% of global population based in cities by 2050 [1, 2]), thus implying and promoting land use changes, not only for the BE growth, but also in relation to agriculture and industry to support population needs. In fact, 1/3rd of the world's mainland surface has changed in the last 60 years [23]. Thus, proper urban development process can be key to tackling the urban heat island (UHI) effect that worsens the heat related SLOD risk in the BE. As mentioned in Chap. 1, Sect. 1.5.1, BE orientation, volume density, aspect ratio (ration between building heights and the facing open space width), sky view factor and materiality (i.e., vegetation and albedo), and their combination, can improve or worsen the microclimatic conditions at the different scales, allowing to allocate hotspots [24], thus having a direct impact on permanent users (e.g., residents) and visitors, especially in outdoors and thus in open spaces.

In particular, exposed open spaces with aspect ratios < 1 are riskier for the only outdoor users (OO), such as passersby, and prevalent outdoor users (PO), such as users of dehors and open terraces, who can populate outdoor areas also during the hours of higher temperatures and solar radiation, and thus in the late morning and the early afternoon. In particular, the heat dynamics impose high stress on users, and air temperatures can grow if wind is blocked, and considerable heat is reflected and/or emitted. Likewise, wind blocking generates low air pollution dispersion and higher air pollutants' concentration for both permanent users and visitors (higher for than for OO and PO). Meanwhile, indoor areas facing the open spaces are riskier especially for residents, as heat is entrapped in the buildings envelope and hardly flushed during nighttime.

Detailed and context-based analysis are useful to determine these characteristics as well as risk mitigation priorities, and the extent to which mitigation strategies are applicable to test how physical vulnerability can be diminished. In this sense, in historical BEs such an issue should be also solved by considering architectural and heritage features, which can limit the possibility to intervene on specific BE features since they are basic parts of the identity of the urban scenario (see Chap. 3, Sect. 3. 4 and Chap. 4, Sect. 4.4).

5.1.2 Users' Exposure

Exposure has to be assessed in terms of space type and time occupation density to provide a complete analysis of users' factors dynamics of the open spaces in the urban BE. Statistically surveyed and maximum capacity of occupation values are present in literature on the differences between building use typologies, as well as outdoor space use or destination (see Chap. 2, Sect. 2.4.2). In addition, when analysing in detail the behaviour of users in public open spaces, their movements within the outdoor areas are highly relevant. The attractors and repulsors of the BE become more significant from a behavioural-based perspective, since they can alter the presence of users over space and time, thus contributing to the possibility to alter the perception of heat stress and then be affected by it since main use areas are more affected by higher universal thermal comfort index (UTCI) levels. Nevertheless, the acceptability of local UTCI conditions depends on the specific activities carried out by users, by distinguishing, at least, between OO and PO [25].

In this sense, analysis on the users' density becomes essential to quantify the exposure, and quick assessment tasks can be based on the intended uses of indoor and outdoor areas. Compared to residential buildings, public open spaces can range up to $\times 10$ the number of people per m^2, although specific differences can exist depending on the considered levels of service of such pedestrian areas. Additional crowding effects implying an increase of exposure can be related to the contribute of use due to the presence of buildings open to the public, and mainly of worship places, museums and cinemas or theatres. These buildings can have a $\times 14$ higher occupation density in respect to residential buildings, thus increasing the number of users who can gather in the immediate outdoors while waiting to enter them.

Nevertheless, exposure varies since occupation peaks can significantly change through the week and through the day in view of the specific intended uses of indoor and outdoor areas. Some relevant peaks seem to occur during weekdays in morning hours (i.e., 11:00–16:00), as shown by Chap. 4, Sect. 4.2. Weather wise, a 1 °C change in UTCI, generates a 4–9% acceptability difference for OO, while a 0.5–4% acceptance difference PO (the higher the UTCI the higher the difference—Chap. 3, Sect. 3.3.1).

Therefore, it is useful to consider in detail the type of outdoor spaces, their surrounding buildings, and the potential users' demographics to better quantify their level of SLOD risk.

5.1.3 Users' Vulnerability and SLODs Effects from a Multi-risk Perspective

Beside the quantification of exposed users in terms of their number (or density), demographic groups and poor health status have to be considered indeed, since they can worsen the SLOD risk for users from an individual standpoint. In particular, heat stress wise, a toddler can be approximately ×6 more prone to suffer heat-related issues than an elder, or ×8 than a male adult. Female youngsters, adults and elders are 0.05–0.07% more prone that the corresponding group of males (see Chap. 4, Sect. 4.3.2). Air pollution distress wise, an elder is more likely to suffer from a cardiovascular failure, it doubles the share of adults,[1] and for someone already prone to cardiovascular issues, the reported relative risk increase is added to their actual health status (see Chap. 4, Sect. 4.3.3).

These results are more significant for toddlers which could be at high risk considering the United Nations' world population growth report. It states that 2/3rd of global population lifetime fertility is below 2.1 births per woman (near the threshold of zero growth in the long run for a population with low mortality). In addition, also that by 2050, elders will be twice the number of toddlers, and about the same as the number under age 12 (children).

In view of the above, considering the overlapping between heat stress and air pollution is essential to provide a proper and comprehensive assessment of SLODs effects from a multi-risk perspective. In this work, the attention is focused on effects on toddlers in view of their paramount fragilities in respect to the rest of the exposed population. According to a conservative approach, the impact of the two separate SLODs could be considered as equal, and this choice could be performed by normalizing the effects of heat stress and air pollution distress, which then vary over a 0–100% scale being perfectly comparable.

5.2 Applications of the "Behavioural-Based" Approach: Insights from Case Studies and Work Perspectives

The contribute to SLODs risk assessment given by the "behavioural-based" approach relies on its experimental-based, user-centered and holistic perspective, and the application to case studies demonstrates the methodology capabilities and robustness in both current scenario analysis and post-retrofit conditions evaluation. In particular, the preliminary application is provided on an experimental-based archetype of a square, as relevant Built Environment Typology (BET) of open spaces in the urban BE as well as on a real-world case study which shares the main BETs features. In particular, both cases consider an historical scenario, referring to the Italian context, with a Cfa—Temperate-hot summer-no dry season climate, a rather enclosed outdoor area (H/W < 1) and an average trafficked street. In addition, to perform full comparisons between the real and archetype scenarios, both of them have same orientation, approximately North–South. Relevant findings related to these applications and referring to the overall risk in the public open space are further described in the following sections, then tracing the future steps in this kind of research and application.

---

[1] https://platform.who.int/mortality/themes/theme-details/topics/topic-details/MDB/cardiovascular-diseases (last accessed: 06/11/2023).

5.2.1 Usability of BETs for Quick Assessment Purposes

In general terms, results on the BET are valid for performing preliminary quantitative SLODs risk analysis with the proposed framework, that can be then better tailored according to the specific real-world case study features. In fact, variance for the overall public open space of the BET and the real-world scenario can be considered null (< 0.006%—Chap. 4, Sect. 4.3.2). This allows to speed up the process of any other virtually modelled BE. In fact, the BET can be used to assess general criticalities due to single and multi-risks, that can be then tailored depending on the specificies of the specific real-world scenario. In this sense, decision makers could be supported while identifying which elements are most at risk in the square, and an inventory of "standard" mitigation strategies could be also derived in the BET contexts through "behavioural-based" simulations.

However, caution should be taken if a BE outside the Italian context is considered [26], and additional comparisons between other archetypes and the related real-world squares sharing the same features have to be performed by future works. Furthermore, any other public open space could be tested, indeed, by following the scenario creation phase (Chap. 3, Sect. 3.2), and supported by literature and open source data repositories for the required inputs.

5.2.2 From Single to Multi-risk Assessment of Heat and Air Pollution

Considering the assessed BET and real-world square (Chap. 4), obtained results raise the alarm for toddlers (other demographic groups and health affection types displayed a much lower risk). Regardless of the type of user (PO or OO) taking care of the toddler, if they remain for more than 2 h exposed to the outdoor environment there is a high risk for the toddler to suffer from heat stroke or heat-related illness (i.e., dehydration).

Air pollution distress did not raise any alarm from the values obtained, although authors acknowledge the limitations on the actual pollution sources inserted in the modelled scenario (Chap. 4, Sect. 4.1).

Mitigation strategies are mainly assessed in this work by focusing on the more critical scenario from a multi-risk perspective (i.e. combining heat stress and air pollution distress effects on toddlers' health), that is the real-world case study. Although their potential impact, only a very limited number of strategies provided little improvement to the current single and multi-risk condition of the public open space. In fact, according to previous works [27], detailed attention is needed while proposing mitigation strategies, as the scale of their intervention can be highly connected to the extend of their impact, and the type of intervention can modify not only the heat stress, but also the air pollution distress.

In the presented case studies (Chap. 4), some of the proposed mitigation resulted in very local effects (e.g., seasonal shadings) which reduced the shortwave radiation effect on UTCI for a single discretized $5 \times 5\,m^2$ area. As a consequence, they have no significant effect on the overall conditions of the studied open space. Other strategies (pavement albedo changes) should be studied at intermediate values that prevent the reflection of solar radiation augment in a higher proportion the heat distress on the PO and OO users. In view of the above, the multi-risk mitigation is not completely reached by the proposed solutions since they seem to be more linked to a single strategy rather than to both of them. Thus, future works should test the mitigation strategies impacts in other ideal (BET) and real-world scenarios, to make sure about the specific impact of an inventory of multi-risk solutions using statistical-based terms. Furthermore, this work does not consider the combination of more strategies, and thus further simulation actions toward these tasks should be performed by using the same proposed methodology.

5.2.3 User's Behaviour Variance in Risk

As expected, users' behaviours in public open space can result in an increase by $\times 1.33$ to $\times 2$ their heat and air pollution distress related SLOD risk (from PO to OO users respectively—see Chap. 4, Sect. 4.3). In particular, for toddlers associated to OO behaviour, the heat stress can be as critical as that their heat-related illness could manifest after 97 min within the studied outdoor areas. Such analysis were performed, herein, using standardized exposure times (i.e. 15 min and 1-h), and thus further analysis could consider the risk with other discretized time steps. Furthermore, the model just considers OO and PO behaviours and standard attraction areas within the open space, but specific point of interests for users could be modelled by considering also temporary uses of the open space itself, e.g. in mass gatherings. The overlapping of additional attractive phenomena due to social interactions with the acceptability of local UTCI conditions could increase the reliability of simulation results, especially in specific case studies. Moreover, the simulation concepts could be better improved by including agent-based modelling issues to detect the movement and position of users over time and space and then applying the equations presented in Chap. 3 from a microscopic standpoint. In this case, the reference grid of particular heat and air pollution stress could be allocated to the single (smaller) discretized outdoor area, although microscopic approaches to environmental conditions simulation should be applied, thus also reducing the current grid dimension (5×5 m). Nevertheless, the overall assessment approach and the specific indicators and metrics will still be maintained, thanks to the demonstration of their reliability.

5.2.4 Risk and Use of Public Open Spaces Due to Heat and Air Pollution

Findings on the behavioural-based approach for SLOD risk also showed interesting insights on how the hazard can considerably modify the way an outdoor space is used. That is, considering the results obtained for mitigation strategies related to surface albedo variation (Chap. 4, Sect. 4.4), the UTCI rises to 43 °C and for B2 48 °C for the application of cool pavement, but the usability of the outdoor area for PO is as low as 10%. Instead, the UTCI rises 46% for permeable pavers, still considering PO only. For OO, these values rise to 68% and 82% respectively. Therefore, the use of the presented behavioural approach allows designers, developers and policy makers to understand not only the risk, but also how the SLOD related risks can slowly affect the public open spaces for common users (residents) and visitors.

References

1. UN-DESA (2019) World urbanization prospects: the 2018 revision. New York
2. UN-Habitat (2022) World cities report envisaging the future of cities. ISBN: 978-92-1-132894-3
3. World Health Organization (WHO) (2014) Quantitative risk assessment of the effects of climate change on selected causes of death, 2030s and 2050s. WHO Press, Geneva
4. World Health Organization (WHO) (2016) Ambient air pollution: a global assessment of exposure and burden of disease. World Health Organization
5. Salvalai G (2021) Rischio dell'ambiente costruito e dei suoi utenti negli SLow Onset Disasters: fattori tipologici di vulnerabilità ed esposizione negli spazi aperti urbani italiani. In: Design and construction tradition and innovation in the practice of architecture (Progetto e Costruzione Tradizione ed innovazione nella pratica dell'architettura). Enrico Sicignano, pp 1446–1462
6. Intergovernmental Panel on Climate Change (2023) Climate change 2021—the physical science basis. Cambridge University Press
7. Metz B (2005) IPCC special report on carbon dioxide capture and storage. Cambridge University Press, Cambridge
8. Jamei E et al (2016) Review on the impact of urban geometry and pedestrian level greening on outdoor thermal comfort. Renew Sustain Energy Rev 54:1002–1017. https://doi.org/10.1016/j.rser.2015.10.104
9. Colaninno N, Morello E (2019) Modelling the impact of green solutions upon the urban heat island phenomenon by means of satellite data. J Phys Conf Ser 1343:012010. https://doi.org/10.1088/1742-6596/1343/1/012010
10. Paolini R et al (2014) Assessment of thermal stress in a street canyon in pedestrian area with or without canopy shading. Energy Procedia 48:1570–1575. https://doi.org/10.1016/j.egypro.2014.02.177
11. Stewart ID, Oke TR (2012) Local climate zones for urban temperature studies. Bull Am Meteorol Soc 93:1879–1900. https://doi.org/10.1175/BAMS-D-11-00019.1
12. Fuladlu K (2018) The effect of rapid urbanization on the physical modification of urban area, pp 1–9. https://doi.org/10.14621/tna
13. Sharifi A (2019) Urban form resilience: a meso-scale analysis. Cities 93:238–252. https://doi.org/10.1016/j.cities.2019.05.010
14. IPCC (2023) AR6 synthesis report: climate change 2023

15. CEM-DAT (2023) EM-DAT: the OFDA/CRED international disaster database. Centre for Research on the Epidemiology of Disasters, Universidad Católica de Lovaina, Bruselas
16. Bernardini G et al (2016) Towards a "behavioural design" approach for seismic risk reduction strategies of buildings and their environment. Saf Sci 86:273–294. https://doi.org/10.1016/j.ssci.2016.03.010
17. Blanco Cadena JD et al (2021) A new approach to assess the built environment risk under the conjunct effect of critical slow onset disasters: a case study in Milan, Italy. Appl Sci 11:1186. https://doi.org/10.3390/app11031186
18. Salvalai G et al (2022) Pedestrian single and multi-risk assessment to SLODs in urban built environment: a mesoscale approach. Sustainability 14:11233. https://doi.org/10.3390/su141811233
19. Blanco Cadena JD (2023) Determining behavioural-based risk to SLODs of urban public open spaces: key performance indicators definition and application on established built environment typological scenarios. Sustain Cities Soc 95:104580. https://doi.org/10.1016/j.scs.2023.104580
20. Curt C (2021) Multirisk: what trends in recent works?—a bibliometric analysis. Sci Total Environ 763:142951. https://doi.org/10.1016/j.scitotenv.2020.142951
21. Han S (2022) Behaviour in public open spaces: a systematic review of studies with quantitative research methods. Build Environ 223:109444. https://doi.org/10.1016/j.buildenv.2022.109444
22. Quagliarini E (2022) Users' vulnerability and exposure in public open spaces (squares): a novel way for accounting them in multi-risk scenarios. Cities 133:104160
23. Winkler K et al (2021) Global land use changes are four times greater than previously estimated. Nat Commun 12:2501. https://doi.org/10.1038/s41467-021-22702-2
24. Huang G et al (2011) Is everyone hot in the city? Spatial pattern of land surface temperatures, land cover and neighborhood socioeconomic characteristics in Baltimore, MD. J Environ Manage 92:1753–1759. https://doi.org/10.1016/j.jenvman.2011.02.006
25. Cheung PK, Jim CY (2019) Improved assessment of outdoor thermal comfort: 1-hour acceptable temperature range. Build Environ 151:303–317. https://doi.org/10.1016/j.buildenv.2019.01.057
26. D'Amico A et al (2021) Built environment typologies prone to risk: a cluster analysis of open spaces in Italian cities. Sustainability 13:9457. https://doi.org/10.3390/su13169457
27. Salvalai G et al (2023) Greenery as a mitigation strategy to urban heat and air pollution: a comparative simulation-based study in a densely built environment. Rev Tema 09. https://doi.org/10.30682/tema090003

Index

A
Acceptability probability (PA), 71
Activities
    leisure, 38
    physical, 37
    social, 38
Air pollution, 2, 7
    background air pollution, 95
    distress, 41
    mortality, 13
Air Quality Index (AQI), 8, 101
Albedo, 24, 94, 109
Alliesthesia, 39
Analysis grid, 70
Aspect ratio, 21
    height and width ratio (H/W ratio), 20
Attractors, 40
    urban attractors, 41

B
Behavioural based approach, 65, 66, 89, 100, 115
Behaviours, 35–37, 40, 44, 122
    adaptive, 42
    simulation, 65, 100
    walking behaviours, 34
Built Environment Typology (BET), 90

C
Carbon emissions, 7
Climate change, 2, 116
    disasters, 9

D
Disability, 33
Dispersion modelling, 69

E
Environmental modelling, 69

F
Familiarity, 97

G
Green
    areas, 20
    walls, 78, 107

H
Heat and thermal stress, 6, 67, 104, 109
    mortality, 12, 72
        Years of life lost (YLL), 72
Heat Index (HI), 6
Heat wave, 4

I
Increasing temperatures, 2
Indoor users, 34, 49, 50
Intended use, 46

K
KPIs assessment, 67, 71

© The Author(s), under exclusive license to Springer Nature Switzerland AG 2024
G. Salvalai et al., *Slow Onset Disasters*,
PoliMI SpringerBriefs, https://doi.org/10.1007/978-3-031-52093-8

M
Micro-climates, 2, 65, 101
Mitigation, 76
　solutions, 79
　strategies, 76, 106, 108
3D Modelling, 69
Multi risk, 116

N
Number of users (NU), 55
　Number of Only Outdoor Users (NOOU), 55
　Number of Prevalent Outdoor Users (NPOU), 55

O
Occupant Load (OL), 47, 50
Outdoor users, 34, 121
　only (OO), 47, 71, 72
　only outdoor (OO), 98
　prevalent outdoor (PO), 47, 71, 98

P
Path selection, 39, 40

R
Relative Risk (RR), 75, 77
Representative Concentration Pathway (RCP), 9, 11
Repulsors, 40
Risk assessment, 17
　exposure, 18, 118
　hazard, 18, 92, 96
　multi risk, 81, 106, 115, 118
　vulnerability, 18
　　physical vulnerability, 18, 20
　　Social vulnerability, 18

S
Scenario creation, 66, 68

Short term risk, 39, 71
　pollution (STPR), 75, 101, 105, 112
SLOD simulation, 2, 70
Slow Onset Disaster (SLOD), 2
　air pollution, 15
Sweat rate, 74

T
Temperature
　air, 3, 5, 6
　humidity solar and wind index (THSW), 6
　humidity wind index (THW), 6
　mean radiant (MRT), 6
　outdoor standard effective temperature (OUT_SET), 6
　universal thermal climate index (UTCI), 6
　wet bulb globe temperature (WBGT), 6
Thermal acceptability, 42
Traffic, 69

U
Universal Thermal Climate Index (UTCI), 101
Urban Heat Island, 3
　atmospheric, 3
　surface, 3

V
Vulnerability
　physical, 116
　Social vulnerabiility, 33, 117
　users, 54

W
Walking speed, 42, 43
Water
　bodies, 20, 107
　loss, 67, 73
　loss risk (WLR), 73, 74